The social basis of
scientific discoveries

The social basis of scientific discoveries

AUGUSTINE BRANNIGAN

Assistant Professor of Sociology
The University of Calgary

CAMBRIDGE UNIVERSITY PRESS

Cambridge
London   New York   New Rochelle
Melbourne   Sydney

Published by the Press Syndicate of the University of Cambridge
The Pitt Building, Trumpington Street, Cambridge CB2 1RP
32 East 57th Street, New York, NY 10022, USA
296 Beaconsfield Parade, Middle Park, Melbourne 3206, Australia

First published 1981

Printed in the United States of America

Library of Congress catalogue card number: 81-6129

British Library cataloguing in publication data

Brannigan, Augustine
The social basis of scientific discoveries.
1. Science – Social aspects
I. Title
306'.4   Q175.S

ISBN 0 521 23695 9 hard covers
ISBN 0 521 28163 6 paperback

Contents

Preface

During the last decade, the sociology of science has undergone a transformation. Not only have traditional perspectives been overthrown, but new forms of analysis have been proposed, new kinds of empirical study have been undertaken, new social groupings have emerged, and increasingly wide-ranging contacts have been established with historians and philosophers. As a result of these intellectual and social changes, the field has often seemed to be in disarray. The old certainties about science, the old belief in its cultural uniqueness and the old landmarks of sociological interpretation have all gone. But it has been far from evident what is to take their place. In my view, the first clear signs of a coherent and viable programme of sociological analysis are now beginning to appear. This book by Gus Brannigan is one of those signs.

The single most important defect of the traditional sociological perspectives on science was that the typical assumptions of scientists themselves were taken for granted by the analysts involved. Thus scientific knowledge was conceived in a straightforward positivist fashion. Scientists' public pronouncements about the values of science were seen as accurately defining the research community's ethical system. Researchers' claim that autonomy was essential for effective knowledge-production was accepted more or less at face value. In other words, scientists' own interpretations of their social world were taken over by sociologists, incorporated into their analytical frameworks and in the process further objectified, that is, made to appear even more 'thinglike' and unquestionable.

In recent years, this objectification of scientists' interpretative achievements by past sociological analysis has begun to be recognized and dismantled in relation to various topics in the sociology of science. Consequently, attempts have been made to devise new and more satisfactory forms of analysis, the aim of which has been increasingly to understand scientists' interpretative practices instead of allowing those practices to dictate the shape of sociological work. Thus, traditionally, sociologists tended to formulate their research topics as follows: given that science exhibits a certain kind of rationality or a type of moral action or whatever, how can these characteristics of science be explained or their consequences identified? In contrast, those working within the newer perspective tend to ask: given that scientists' actions and beliefs can be construed in various ways, by what interpretative practices is science made to appear to exemplify a certain kind of rationality or a type of moral action or whatever?

Until the publication of the present book, however, this kind of revision of traditional questions has not been attempted in relation to the topic of discovery, with the exception of a single paper by Woolgar (see chapter 5 below). This is surprising because, as Brannigan points out, the notion of discovery is prominent in our conception of science. Thus this book fills an important gap in the current reconceptualization of the sociology of science. Moreover, because there is virtually no earlier work on which to build, the author has had to do it all himself more or less from scratch. Accordingly, he presents a systematic critique of existing theories of discovery. He erects on this critical evaluation an alternative analytical perspective. He illustrates the fruitfulness of this perspective in the interpretation of specific examples of discovery. He offers a highly original account of the origins of existing theories. And he identifies some of the lines of research which are prompted by his 'attributional theory of discovery'.

Brannigan suggests that previous theories of discovery are of two types: those which rely on mentalist notions, such as 'genius' or 'gestalt switch', and those which employ a conception of 'cultural determination'. These two kinds of theory have in common the assumption that discoveries are naturally occurring events which can be identified by the analyst without too much trouble and the assumption that, once identified, discoveries can be explained by linking them to some preceding event. Brannigan challenges both these assumptions. He demonstrates that discoveries as such are not naturally occurring events. They are 'events' whose status as discoveries is variable and dependent on the contingent interpretations of participants. Previous analysts, in identifying discoveries, have either adopted these lay interpretations ready-made or have themselves engaged in the same kind of contingent interpretation as they sort through conflicting versions of events to achieve a 'correct categorization'. Brannigan also shows clearly that the attempts at explanation are uniformly unsuccessful and that the theories are, on the whole, analytically empty. In many cases, they are mere tautologies. For instance, the only evidence of 'genius', as distinct from high intelligence, is usually just that discovery which is to be explained. In addition, these theories are often unable to distinguish between discoveries and situations of ordinary learning, that is, situations where an individual comes to understand for the first time some conception which is already widely understood by others.

The author seeks to avoid the inadequacies of traditional theories by focusing on the social context of discovery rather than on the mental event of discovery. He stresses that a theory of discovery must deal specifically with the social category 'discovery'. It must distinguish between discovery and other possible categorizations, such as learning, replication, plagiarism, presentation of the self-evident, fraud, fantasy, and so on. This can only be done by taking into account the social context in which actions and mental experiences occur and as a result of which various categorizations are applied. The theory must also be able to deal with changes in the attribution of categories to particular actions or knowledge-claims; that is, we have to recognize that what are now known as discoveries were not necessarily always so defined and that what were previously said to be discoveries are not always treated as such today. Thus, the theory must come to terms with the fact that the categorization of a scientific contribution as a discovery is an interpretative and variable accomplishment on the part of those actors who create the social contexts in question.

Brannigan concludes, then, that discoveries do not occur as discrete, causally explicable, events. 'Discoveries' are rather context-dependent categorizations achieved by participants in pursuit of their own practical objectives and 'explicable', for those practical purposes, in terms of participants' own lay theories. Traditional theorists have made the mistake, not only of treating participants' contingent identifications of discoveries as literal descriptions, but they have also taken over scientists' lay theories. Unfortunately, as Brannigan shows, these folk theories, although they are entirely appropriate for and satisfactory within the requirements of everyday informal scientific discourse, are quite unsuitable for the purposes of systematic sociological analysis.

In the central chapters of this book, Brannigan demonstrates how sociologists can begin to take as their topic for analysis scientists' theories about, and procedures for constituting, discoveries. He shows how, instead of taking discoveries for granted and trying to account for them, we can pursue the prior issue of how certain events are made out to be discoveries in the first place. His procedure is to explore the common sense meaning of 'discovery' and to identify the basic criteria through which discoveries are defined, recognized and constituted by participants. He identifies criteria dealing with four aspects of discovery; namely, the feasibility of a knowledge-claim, its validity, the kind of motivation involved, and the degree of originality. It is these elements of an act, Brannigan suggests, which scientists, and laymen, take into consideration when deciding on the attribution of the term 'discovery'. This interpretative machinery for the constitution of discoveries is then used by the author to illuminate the chequered history of Mendel's claims, the Piltdown Man, and so on, in order to show dynamically how the status of these claims as discoveries depended on members' variable employment of this machinery.

Brannigan himself does not use the phrase 'interpretative machinery'; probably because it might seem to imply, misleadingly, that participants' procedures for construing discoveries are rigidly formalized. However, the mechanical metaphor *is* helpful insofar as it draws attention to the objectification of the interpretations which members of society produce. For once participants construe an act as a discovery, the act comes routinely to be described in terms of the category 'discovery' and to be embedded in lay theories of discovery. To refer to the act at all is to refer to a *discovery* and to draw upon the interpretative vocabulary appropriate to a discovery. The fact that it was a discovery comes to *appear to be* a natural fact of life.

I will not try to convey any more of the content of the book. It is impossible to do it justice here. But it is worth commenting briefly on two of its possible implications. Brannigan shows to my satisfaction that 'discovery' can best be approached, sociologically, as a method or interpretative practice which scientists use in making sense of the events going on around them. His formal criteria can therefore be seen as a set of rules or procedures for construing events as discoveries. However, he also shows clearly that, although members of society appear to employ similar rules, they often reach quite different conclusions about the character of particular knowledge-claims; and he describes in detail how scientists came to divergent conclusions in various specific instances. Furthermore, we know from previous work that the meaning of a set of rules is never self-evident in its application to particular cases and that rules always require further interpretation which cannot be fully specified by the rules themselves. Now it is possible that the varying implementations of the procedures for constituting discoveries are due to purely local, idiosyncratic features. In which case, if we assume that Brannigan's substantive analysis is broadly correct, the sociological study of discovery would have little further to go conceptually. It would consist primarily of various illustrations of how Brannigan's machinery was put into operation on specific occasions. The other possibility, however, is that these 'surface rules' are linked to one or more repertoires of supplementary interpretative procedures, which make possible a range of specific interpretations in any one.case, yet which can be formulated analytically. If this is so, and it seems to me to be the more likely possibility, the author's analysis will have opened up a whole new area of sociological research.

The second issue to which I wish to draw attention is whether it follows from Brannigan's analysis that discovery is a purely linguistic phenomenon. The author himself recognizes the importance of this issue and he states clearly that this conclusion does not follow necessarily from his argument. He writes that his four criteria constitute a set of necessary and sufficient conditions for the definition of an act or a knowledge-claim as a discovery and that 'the criteria and their use explain members' action regarding new theories'. Thus the idea is that, once we have understood how an act comes to be deemed to be a discovery, this will help us to explain why certain kinds of *action* then ensue. The problem, of course, is how can we characterize these ensuing actions without adopting in our treatment of the consequences of discoveries just that form of analysis which has been abandoned in relation to discovery itself? Once the analyst has withdrawn from the analysis of discovery as action, and replaced it with an examination of discovery as a participants' method for construing action, he seems to be obliged to treat all other classes of action in the same way. For there is no realm of social action where it is possible to identify 'what really happened' without treating at least some part of members' interpretations of what happened as analytically unproblematic, literal accounts.

It seems to me, therefore, that the main thrust of Brannigan's study is towards an analysis of 'discourse about action' rather than of 'discourse *and* action'. It also seems to me that this is eminently sensible. For it is scientists' discourse, that is, their documents, recorded utterances and pictorial representations, which is available to us for study. Thus, for me, Brannigan's book is a major step away from the traditional and unattainable objective of using participants' statements as the basis for definitive characterizations and explanations of 'what really happens' in science, to a more realistic concern with the various kinds of accessible text in which are embodied the practices whereby scientists give meaning to their world. Not all readers will want to follow me that far. But even so, few readers will deny that this book significantly revises the sociological approach to the study of discovery and that it leads us in an intriguing new direction.

MICHAEL MULKAY

Princeton University and
University of York
October 1980

I

The topic of discovery and the concept of nature

The topic of discovery dominates the imagination of scientists in their working lives as well as that of students of science in their studies; as N. R. Hanson notes, 'discovery is what science is all about'.[1] It would appear that the primacy of discovery is derived from the ambitions of the field from which it arises. Science, that peculiar culture which is the hallmark of Western civilization,[2] makes the discovery or uncovering of nature its central focus. This conception of scientific activity as the dispelling of the unknown presumes a distinctive mystery or intrigue in the notion of nature which pervades the work of science. For example, Galileo's conception of the *mathesis universalis,* the 'underlying' relationships which govern nature, overshadowed the world as it was known directly to the senses, and challenged the common sense world of the Ptolemaists.[3] Man in Protagorean terms was no longer the 'measure of existence'. In the *Theatatus,* Socrates had spoken of Protagorus' theorum that 'man is the measure of all things: alike of the being of things that are, and the not-being of things that are not'.[4] Socrates spoke of the Protagorean concept of truth as the 'aisthesis', the directly knowable world – we have translated this as 'sense perception'. Centuries later, Galileo again raised the spectre of 'sense perception'; he claimed that the senses were 'subjective', and hence were untrustworthy and unreliable. The true field of nature was something out of the ordinary, and consequently the scientists explored a quasi-physical or metaphysical realm to determine the 'structure' or 'form' of the laws which underlay the directly perceived world. Common sense was no longer valid, and traditional thought about the world became the subject of studied scepticism.

Similarly, Descartes' theological efforts to provide a solid basis for scientific knowledge in his *Discourse on Method* were far from our demystified modern materialism. In Descartes' epochal dualism, the guarantee of the correspondence between what was impossible to doubt 'internally' and what actually existed 'externally' in nature was provided by God.[5] The godhead, that mystery of mysteries, simultaneously became the foundation for the self on the one hand, and for science on the other. One finds an uncanny reverberation of this thinking in Newton's picture of space as 'God's mind'. All these images lent a mysterious aura to nature and reified the process of discovery. This direction in early scientific culture had certain discrete consequences. The world as it was known in traditional and common sense ways became distrusted. Science directed its attention to a quasi-physical 'nature' or underlying order of things which had a characteristic intrigue associated with it. And because of the mystery associated with nature, the procedure of its becoming known came to exhibit a dramatic social significance. Consequently we find a curious feature in accounts of scientific discoveries; they are recurrently characterized as being bizarre achievements made by eccentric personalities under curious circumstances, often having horrible consequences. For example, in the traditional myth, Daedalus escapes his captivity by affixing feathers to his arms with wax, but falls to his death as the wax melts when he approaches too close to the sun. Midas faces an equally unhappy fate when his power to change things to gold makes life impossible. A whole series of dystopian novels like Aldous Huxley's *Brave New World,* Zamiatin's *We,* and Samuel Butler's *Erewhon* depict in various ways the ironic turns for the worse that follow the advance of science.[6]

As for the bizarre depictions of the process itself, the iconic image of discovery is provided in classical accounts of Archimedes who, naked and distracted, is said to have run from the gymnasium baths proclaiming 'eureka' after having discovered the laws of hydrostatic displacement. Presumably this was not what Hans Reichenbach had in mind when he spoke of justification as a method of 'presenting [a discovery] before a public'.[7] The history of science is filled with accounts similar to that of Archimedes. Alfred Russell Wallace, recovering from fever during his research in the dense jungles of the Malay Archipelago, reported that he was struck by the idea of speciation *in a delirium* and proceeded to write it out in a single sitting.[8] Einstein reported to Michael Polanyi that the idea of general relativity became vivid to him as a result of a youthful dream in which he tried to follow a beam of light.[9] Kekulé relates that the hexagonal structure of benzene molecules became apparent to him whilst staring half asleep into his fireplace; imagining the flames were snakes, he saw one bite its own tail, forming the hexagonal ring – the very form he was looking for.[10] And then there is the familiar tale of Newton and the apple.

So in the popular images of science, the kernel of scientific thinking is often shrouded in a shell of mystery and/or irrationality, in a dream, or in a fit of distraction, or in the eccentricity of a historical personality.

These conceptions of science are vividly confirmed in the literary images of Dr Faustus, Dr Frankenstein and Dr Jekyll. The scientist and the act of discovery are repeatedly represented as sources of intrigue and mystery, associated with the bizarre and the irrational. It is my conjecture that this characterization of discovery may derive from the shift of attention from the scholastic 'world' of the middle ages to the unknown 'nature' of the Renaissance, from the world of common sense knowledge and traditional belief, to the mathematical nature of existence. This shift is nicely reflected in the change of attitude regarding the formal representation of the world. In Cusanus, the arithmetical models of nature are referred to as 'De Conjectura' – conjectures. With Galileo, the shoe is on the other foot: the real world is the 'mathesis universalis', and the world of everyday life is elusive and 'conjectural'.[11]

REICHENBACH'S DISTINCTION: DISCOVERY VERSUS JUSTIFICATION

The philosophical study of science in the twentieth century appears to have avoided the popular images of scientific discovery and the ostensively irrational aspects of it under a directive from the 'positivist' movement. Hans Reichenbach suggested that the *actual thought processes* and historical conditions whereby a new law or a new mathematical demonstration is arrived at are different from the *rational reconstructions* which occur when the scientist or mathematician communicates the new theory to others. This supposition has become elevated to the status of a *doctrine* of the separation of the context of discovery from the context of justification. However, having made such a principled distinction, the philosophers have generally assumed that *only* justification could be amenable to logical analysis; hence the context of discovery had no status as a philosophical problem. Thus Karl Popper, in a book whose title was mistranslated as *The Logic of Scientific Discovery*, legislated the problem out of philosophy by relegating the matter to 'empirical psychology':

The initial stage, the act of conceiving or inventing a theory, seems to me neither to call for logical analysis nor to be susceptible of it. The question how it happens that a new idea occurs to a man – whether it is a musical theme, a dramatic conflict, or a scientific theory – may be of great interest to empirical psychology; but it is irrelevant to the logical analysis of scientific knowledge.

There is no such thing as a logical method of having new ideas, or a logical reconstruction of this process.[12]

Popper's entire work, which was originally published in 1934 as *Logik der Forschung* (i.e. Logic of Research), is concerned with the formalization and falsifiability of hypotheses. Popper's view re-articulates the position originally formulated by Hans Reichenbach in 1930, in *Erkenntnis,* the journal for the movement of the unity of science. Though Reichenbach reiterates the distinction in several of his other works which are available in English, the original article, to my knowledge, has never been translated. The grip which Reichenbach's doctrine has had on modern authors is due to the influence of Popper. However, it is by no means limited just to his writing. Richard Braithwaite in his *Scientific Explanation* reiterates this opinion: 'The solution of these historical problems involves the individual psychology of thinking and the sociology of thought. None of these questions are our business here.'[13]

Hence the philosophy of science was seen to concern itself entirely with the function of objective arguments in the 'justification of change in science',[14] that is, with the construction of precise hypotheses which could be presented in favour of an idea, the observations which these hypotheses illuminate, and the results of the attempts to falsify them.

Specifically, for Karl Popper the topic of the historical processes of discovery was displaced by the conception of scientific theorizing as the formulation of conjectures about nature. Hence the model of *Conjectures and Refutations* was not necessarily a description of how scientific thinking was actually done, but was a normative idealization, or an injunction about how it ought to be conducted. R. G. A. Dolby touches on this point:[15]

Nineteenth-century writings on scientific method claimed to be describing the process which successful scientists actually used, and which all scientists *ought* to use. There was no divergence between history and philosophy of science. But with the rise of Logical Empiricism, the descriptive claims of the nineteenth century seemed to be abandoned. With philosophers like R. Carnap, no attempt was made to reflect the activities of practising scientists. The explicit motive of the methodologist was to set out an ideal that scientists should *aspire* to follow ... It is not difficult to demonstrate the separation between the logic of scientific method ... and the actual scientific procedures revealed by historical study.

This movement in the study of science had two prominent consequences. First of all, though writers like Popper and Reichenbach may have had good reasons for separating the actual behaviour which resulted in the discovery of new laws from the subsequent presentation and/or demonstration of their validity, the confinement of attention to justification had an unhappy consequence: it inadvertently contributed to the popular notion that the act of discovery was some mysterious process. As Richard Blackwell later noted, Popper's action obscured the problem by laying it to rest in psychology, a 'convenient dustbin' for philosophical problems. One imagines that Reichenbach and Popper reeled back from the image of Archimedes soaking wet and distracted, and directed their attention to the reasons which demonstrate the hydrostatic laws. Only later did philosophers give serious attention to a theory of the actual *in situ* behaviour.

The second major consequence was that the Reichenbach distinction created a rift between the normative picture of scientific theory construction which was offered by logical empiricists on the basis of finished theories, and the descriptions of scientific practice by historians and behavioural scientists based on studies of how research had actually been conducted. An illustration of this rift is offered by the reaction to Thomas Kuhn's accounts of the way in which historical changes in science had been brought about. Many students of the 'reconstruction' school were quite shocked by Kuhn's model of world-outlook shifts, for they appeared to portray the adoption of new theories as a type of 'mob psychology'. One writer characterized Kuhn's work as a 'Frankenstein, which no amount of reformulation can call back'.[16] Though I believe these reactions to Kuhn are indefensible, they underline one of the great costs paid by the Reichenbach distinction: it produced a tension between scientific *practice* as a topic and finished scientific *theories* as a topic. Furthermore, it appears to have created the *impression* that an account of the 'logic' of scientific discovery, i.e. a demonstration of a theory's validity, was simultaneously an account of scientific discovery, i.e. *how* the idea had occurred to an individual, so that for many people, Popper's *Logic of Scientific Discovery* was read as an account both of the process of discovery and of the validation of theories.

Not only behavioural scientists but their objects – the natural and physical scientists – were aware of the rift between the actual practices of inquiry and the normative idealizations of the philosophers. For example, in reply to Reichenbach's logical axiomatization of relativity theory, Einstein responded that 'he did not find it convincing even on its own grounds'.[17] Reichenbach further attempted to reconstruct Einstein's actual inferences as though they were the result of a 'radical empiricism in a field which had always been regarded as a reservation for the discoveries of pure reason';[18] in other words, Reichenbach appeared to equate the actual process of inference with his formalized reconstruction. This too met with Einstein's objections.

Holton notes, in reply to Reichenbach's essay: 'Einstein devoted most of his attention to a denial of this claim.'[19] Clearly, the representation of the logic and/or rational reconstruction of relativity theory created a picture which was unfamiliar even to its originator. However, given Reichenbach's doctrine, this infidelity was not unjustified; strictly speaking, philosophy is only interested in the completed theory and the grounds of its validity, not in its fidelity to the historical process or its familiarity to the originator.

There were two reactions to these consequences of the Reichenbach distinction. The first took the form of an attempt to describe various 'logics of discovery' based on the *in situ* reasoning of scientists in their actual research. These logics constitute theories of discovery, that is, accounts of the processes which result in discoveries. This book is directed largely to an evaluation of such efforts. However, there has been a second reaction: many authors have rejected the relevance and validity of the distinction. As we shall see, these latter efforts are not theories of how discoveries occur, but theories of how discoveries do *not* occur. We shall deal briefly with this latter reaction first.

REICHENBACH'S DISTINCTION AND THE HISTORICAL RECORD

Logics of discovery are not intended to undermine the distinction between the context of discovery and justification but to supplement what is known of the latter with a description of the logic of the former. Such efforts, from the point of view of philosophy, reclaim ground that was hastily abandoned by the early positive philosophers. The work of Feyerabend, Holton and Kuhn, on the contrary, challenges the integrity of the distinction itself.

The best illustration of this position is the work of Paul K. Feyerabend. Feyerabend's analysis of Galileo's theories indicates that Galileo brought about allegiance to the Copernican system through the deceptive use of new natural interpretations of motion which, unknown to his Aristotelian opponents, concealed a highly abstract observational language. This was subsequently 'justified' by Galileo's self-serving interpretations of telescopic images which, though contradicted by what could be seen with the naked eye, supported the Copernican view. Consequently, *his justification and his discoveries could hardly be said to be independent.* His justification also had other dimensions: Feyerabend argues that

Galileo prevails because of his style and his clever techniques of persuasion, because he writes in Italian rather than Latin, and because he appeals to people who are temperamentally opposed to the old ideas and the standards of learning connected with them.[20]

It is clear that allegiance to the new ideas will . . . be brought about by means other than arguments. It will . . . be brought about *by irrational means* such as propaganda, emotion, *ad hoc* hypotheses, and appeal to prejudices of all kinds. We need these 'irrational means' in order to uphold what is nothing but a blind faith,[21] . . . an unfinished and absurd hypothesis . . .[22]

Far from deploring this state of affairs, Feyerabend recommends it.

What our historical examples seem to show is this: there are situations when our most liberal judgments . . . would have eliminated an idea or a point of view which we regard today as essential for science . . . The ideas survived and they can *now* be said to be in agreement with reason. They survived because prejudice, passion, conceit, errors, sheer pigheadedness, in short all the errors which characterize the context of discovery, *opposed* the dictates of reason . . . *Copernicanism and other 'rational' views exist today only because reason was overruled at some time in their past* . . . Hence it is advisable to let one's inclinations go against reason in any circumstances, for science may profit from it.[23]

Feyerabend concludes that

The results contained so far suggest abolishing the distinction between a context of discovery and a context of justification and disregarding the related distinction between observational terms and theoretical terms. Neither distinction plays an important role in scientific practice. Attempts to enforce them would have disastrous consequences.[24]

A determined application of the methods of criticism and proof which are said to belong to the context of justification, would wipe out science as we know it – and would never have permitted it to arise.[25]

Reichenbach's distinction is unfounded, according to Feyerabend, because real conceptual advances in science transform the very criteria of justification, i.e. observations and proof. For example, Galileo's belief in the reliability of the telescope was co-emergent with its 'confirmation' of his Copernicanism. Ludovico Geymonat writes, 'In Galileo's mind, faith in the reliability of the telescope and recognition of its importance . . . were two aspects of the same process.'[26] This problem which has been called the theory-loaded character[27] of data also informs Kuhn's rejection of the Reichenbach doctrine. Notes Kuhn, 'My attempts to apply [these distinctions] even *grosso modo* to the actual situations in which knowledge is gathered, accepted and assimilated have made them seem extraordinarily problematic.'[28] They have been problematic because the radically new theories transform observational terms and objects simultaneously with their theoretical counterparts! In other words, justification and discovery occur simultaneously.

Gerald Holton's view is quite similar. Holton argues that *in addition* to the logical and empirical aspects of explanations there is a third element: the themata.[29] Themata are pre-theoretical suppositions about nature: for example, that it is mathematically harmonious, that it is composed of fundamental units, or atoms, that it is mechanically integrated like a clock, that natural forms are symmetrical, inherently aesthetic, etc. Holton notes that the major consideration for Einstein's famous paper was not the work of Lorentz or Michelson, but the aesthetically disturbing *asymmetry* of Maxwell's equations. Einstein's alternative account had an inner consistency or symmetry which Maxwell's did not. It was this thematum which recommended the theory, in spite of the fact that the first response to the paper to appear in the scientific community was 'a categorical experimental disproof of the theory'. Nevertheless, and before the confirming experimental data appeared, physicists endorsed the theory because of its *thematic* element. Thus Wilhelm Wien wrote: 'What speaks for it most of all . . . is the inner-consistency which makes it possible to lay a foundation having no self-contradictions, one that applies to the totality of the physical appearances . . .'[30] Nor was Einstein prepared to be discouraged by later empirical disconfirmation. Eddington's observations in 1919 clearly confirmed Einstein's predictions about the deflection of starlight by solar gravity. However, Einstein suggested that even if the results had been negative, 'then I would have been sorry for the dear Lord – the theory is correct'.[31]

This only reinforces the position articulated by Feyerabend: that certain of the greatest historical discoveries were not in accord with Reichenbach's distinction, nor could they be, given the theory-ladenness of observations, nor *should* they have been, had that been possible. For example, if Einstein had followed Reichenbach's protocol, he would have recanted his views in 1905 following the experimental disproof, and would never have tolerated the post hoc justification of relativity by Michelson's unrelated experiments of 1887. Indeed, Einstein himself did not become aware of the ether experiments until *after* 1905. Apparently, compliance with canons of logical empiricism would have retarded the whole revolution of twentieth-century physics which Einstein initiated. The same applies to Galileo, according to Feyerabend; a strict experimentalist would have stayed with a Ptolemaic cosmos.

These accounts of scientific discoveries do not constitute a theory of how discoveries occur. They suggest on the contrary how they have *not* occurred: by following the method of conjecture and refutation, and carefully discriminating between elements which suggest a hypothesis and those which justify it. What is recommended by Feyerabend *et al.* is a new epistemological position regarding research and discovery: Feyerabend's conclusion is simply and literally 'anything goes'.[32] Holton's work has a somewhat related intention: 'to prompt the educator to re-examine conventional concepts of education in science',[33] by paying full heed to those thematic aspects of explanations in science which have guided important discoveries, but which have been obliterated by the inaccurate reconstructions found in the science textbooks. In other words, Holton and Feyerabend's conclusions are *prescriptive;* on the basis of how science has actually operated, they advise how texts should be written and how research should or, more correctly, should *not* be conducted. Though of interest to the discussion of Reichenbach's doctrine, that conclusion is not of immediate importance in this work, nor are any epistemological studies which argue which theory of knowledge the scientist should adopt. These are more clearly the subject matter for the philosophy of science. Our task is different: we are concerned with models of how discoveries have actually occurred. Consequently, our domain is behavioural, not philosophical. Our major concern with discovery is: how has it occurred, and can we describe such occurrences with an adequate theory? This is the second notable reaction to Reichenbach's distinction.

The logic of discovery and the action of scientists

Students of science have frequently investigated the context of discovery under the rubric of 'the logic' of discovery. Presumably this provides a pleasing symmetry with 'the logic' of justification. However, these terms refer to two qualitatively different realms. The logic of justification refers to the definitional coherence of, and relations between, specific variables in a constructed model. These are logical and empirical matters par excellence. However, when we speak of 'the logic' of discovery, this is hyperbolous. Only when an account of such action has been shown valid will we be able to speak unambiguously of the logic of *the theory* of discovery (i.e. 'the logic of discovery'), meaning the formal and/or empirical reliability and validity of the theory of this type of action. Consequently, this usage is ambiguous. The topic it refers to is the research activities associated with the production of discoveries, and the chief question it pursues is what are the conditions which produce or control the occurrence of these discoveries.

When we approach the problem in this way, we are doing a number of things. We are making the topic an examinable or 'researchable' phenomenon that begins with a mindfulness of the mystical notions found in folklore. Also, we are separating the problem of the theory of discovery from the question of whether the *in situ* logic, whatever it is, is *different* from its retrospective reconstruction; that was Reichenbach's phenomenon. Consequently, our interests lie neither in the question of how it *ought* to occur nor in the question of how it does *not* occur. How it *ought* to occur is a matter for normative-minded methodologists; how it does *not* occur has been a matter for a group disputing the claims of the former. There has been a certain futility in this latter debate. When we consider that the methodologist is recommending an idealized prescription for research outcomes, the claims by others that discoveries have not in fact occurred in this way is, strictly speaking, addressing a different set of concerns. The anti-Reichenbachian position that recommends 'anything goes' can only be seen to be answering the methodologist's prescriptive plans if the latter can be heard to confuse his *pre*scriptions of how science should be done (based on hindsight) with *de*scriptions of how it has actually happened. We have seen that Reichenbach does indeed confuse these issues when he represents Einstein's actual reasoning in terms of his own reconstructed axiomatization of relativity. Having recognized this confusion, we see in our own minds that some of Reichenbach's critics are not simply postulating an alternative model of how scientists have made discoveries – they are recommending an alternative epistemology or methodology. That is, by raising the question of how discoveries *have not* been discovered, they are not constructing a model of discovery, but are opening up an *alternative* model of how research *ought* to be done. This, however, is not our question. How things 'ought' to be done is a separate topic from how things have been done.

Having clarified the problem this way, we see how this work is capable of responding to the two consequences of Reichenbach's doctrine outlined earlier. The problem is no longer 'discarded' by relegating it to psychology; indeed we shall spend a good deal of time investigating psychological contributions to the question. Furthermore, having clarified the nature of Reichenbach's distinction, we see that there is no overlap between the philosophical problem of explicating the formal and empirical adequacy of reconstructed truth claims about nature, and the behavioural science problem of explaining the conditions under which certain social phenomena, namely discoveries, occur. These are problems for separate domains.

NATURALISTIC AND ATTRIBUTIONAL APPROACHES

When we pose the question, how did some particular discovery occur, there are at least two possible *types* of answers that can be offered. In the next chapters, I will outline what I take to be the most prominent types of explanations which account for the occurrences of discoveries. These explanations are predominantly mentalistic. That is, they treat discoveries as the outcome of some earlier mental condition or event which has, in this respect, a specific causal value. For this reason I call such accounts 'naturalistic' in that they posit discovery as a type of natural phenomenon whose occurrence is controlled by the presence of certain other phenomena to which it is nomothetically or inherently linked. There is a range of psychological mechanisms which have been discussed in this way by numerous prominent authors. There is a second naturalistic explanation which arose historically in reaction against the first. This treats discoveries as a function of cultural growth, and gives only contingent status to mentalistic conditions. This account, in that it treats discovery in terms of a historically causal matrix of events, is also classified as 'naturalistic'.

In the following chapters I offer an analysis of these models which inspects their formal adequacy and empirical relevance, and compares them to an alternative attributional model. This model defines the problem of discovery somewhat differently: it suggests that we should explain how certain achievements in science are *constituted* as discoveries – and not how they occurred to an individual. This model focuses on how persons confer the status of 'discovery' on social events and how they determine and sanction the appropriateness of this category both for their own achievements and for those of others.

2

Psychological accounts of discovery

It is my contention that models of discovery in the current literature are almost uniformly *psychological* in orientation. By this I mean that they equate the task of explaining discovery with the task of explaining how an idea gets into an individual's mind. Furthermore, all these psychological or mentalistic models tend to employ the same concept or mechanism in their explanations. This mechanism is called various things and is employed in slightly different ways; however, in my estimation the mechanism found throughout the literature is some variant of *gestalt shift*. In this chapter I would like to describe in detail the explanations of discovery offered by several prominent writers: Norwood Russell Hanson and Richard Blackwell, Thomas S. Kuhn, and Arthur Koestler. These writers represent their works respectively as a philosophical approach, an historical approach and a psychological approach. However, I will endeavour to illustrate the central common elements which constitute the explanatory kernel of each. This is not at all an impossible task in that all these writers, in spite of their orientations, appear to be motivated by the same object, and to be seeking the same end; specifically, they all appear to reject the doctrine of Reichenbach and seek to describe the ways in which scientists actually have made historical discoveries. Consequently, whatever their discipline, each is proposing a model of discovery, i.e. an explanation of the conditions which bring about discoveries.

I should also point out that these writers are only representatives of a much larger class of writers. My strategy here is not to be exhaustive; but to represent a *kind* of explanation, and to describe its strengths and limitations. Though I will probably try the reader's patience with the chapter even as it stands, it could be substantially longer if space were devoted to writers like Wertheimer, Polanyi, Poincaré, Taton, Hadamard, Peirce, etc. However, I expect that the arguments I raise in the works examined here will apply likewise to the entire gamut of such *types* of explanations.

Lastly, I would add that while I am at times very critical of certain writers, this should not be read to mean that I find their work entirely without value. On the contrary, all the explanations examined here have been generally acknowledged to be quite valuable. My point will be that there are certain costs to be paid in limiting our attention to these kinds of explanations. I, of course, will offer an alternative account which will inevitably have its own benefits and liabilities. Also, the perceptive reader will see that, though I challenge the thrust of these approaches, I have borrowed various ideas from each of the writers in the construction of my own model. We turn now to the works under consideration.

HANSON AND THE LOGIC OF INFERENCE

Norwood Russell Hanson has been the first modern philosopher to insist, contrary to Popper, on the investigation of the logical grounds of discovery. Indeed, no one objected more strenuously to the delegation of the matter to the behavioural sciences than he. While he acknowledged the relevance of the psychological processes of 'intuition' and 'insight', he argued that the spectacular reorganization of concepts associated with such processes were of 'profound epistemological importance' and could be examined empirically.[1]

Hanson's efforts to spell out this process of discovery emerged in three separate studies: *Patterns of Discovery;* 'Is there a logic of scientific discovery?'; and 'Retroductive inference'.[2]

The central thesis of *Patterns of Discovery,* in its most condensed form, is that the identity of phenomena is a function of their conceptual background. Hence the factual character of observations, of causal linkages, even of so called 'brute facts' emerges only in a contextual placement. In anticipation of Kuhn, Hanson rejects the sense perception theory implied by enumerative induction, and focuses on the gestalt organization of perception. For example, we can recognize a familiar melody – even though it is played in a different key and even though the timing has changed – because we can grasp the overall organization governing the single notes. Likewise, discovery is understood as the reorganization of the value of an object or fact or pattern, in terms of a more comprehensive system of relations by which it is seen to be defined and circumscribed. This idea anticipates Kuhn's concept of the 'paradigm' and the unique gestalt by which the facts, observations and patterns of relations are anticipated and 'conjugated'. In this regard, Hanson raises the prominent example of the 'incommensurable' visions of Brahe and Kepler in the light of the dawn; for the one, the morning *sun was fixed* while the earth moved away in its orbit. For the other, the *earth was fixed* while the sun moved.[3] The visual 'data' supported either view. What was definitive was the conceptual organization by which both observers interpreted the event. According to Hanson, the discoverer is 'the man who sees in familiar objects what no one else has seen before'.[4]

In the article 'Is there a logic of scientific discovery?',[5] Hanson reformulated the general relevance of contextual organization into a logical specification; the logic of discovery concerned the relevance for the scientist of different types or classes of hypotheses over other types. That is, at the outset the scientist will have reasons for preferring certain candidates or conjectures over others (in accord with his existing conceptual inventory), and from these he will seek the specifically adequate hypothesis. In other words, scientists do not conjecture wildly or blindly in their research. For Hanson, the initial postulation or conjecture already involves a logical step of discriminating between promising versus implausible categories of hypotheses.

In 'Retroductive inference', Hanson outlines the process whereby such discrimination of types of hypotheses hinges on the initial observations of anomalies.[6] As Hanson notes, this involves an exact reversal of the process of inference involved in hypothetico-deductive reasoning. In *finished* theories, low level observations are shown to be deduced from high level postulates through reasoned correspondence rules and inference steps. Hanson claims that something resembling the same inference network must have been experienced by the scientist who reasoned inductively from a striking case back 'up to' the general law.

The logic of discovery in this instance is the set of inferential steps whereby an anomaly, or low level observation, is used to generate the context or gestalt by which it is furnished. The overall pattern of discovery which Hanson synthesized in his three papers together provides the following model:

A Logic of Discovery should concern itself with the scientists' actual reasoning which
    a. proceeds inductively, *from an anomaly,* to
    b. the delineation of *a kind of* explanatory hypothesis which
    c. fits into an organized *pattern* of concepts.[7]

Hence the logic of discovery is the process of 'the actual reasoning' whereby the researcher moves from stage to stage in an argument, 'even before a conclusion is reached and tested'. Hanson's assumption here is that the logic of discovery is a form of inductive reasoning engaged in by the scientist in his search for solutions to perceived puzzles or anomalies. By definition, this logic is the pattern of inference which – once the discovery has been made or announced – is seen to have temporally anticipated the final breakthrough or solution.

Discovery versus learning in Hanson

Is this description adequate for the understanding of discovery as such? It seems initially that Hanson is offering a very appreciable grasp of the nature of discovery. However, let me suggest that by treating the logic of discovery at the outset as a variety of inferential logic conducted by the researcher, Hanson leaves no provision in the analysis for a distinction between discovery and *learning*. In other words, Hanson's model, in that it confines its examination to the initial role of anomalies or puzzles, the discrimination of promising leads (or hypotheses), and the gestalt reorganization of the subject domain, is as equally applicable to the research scientist as to the rat in the maze, or the cook whose cakes won't rise. In other words, under Hanson's model, *the logic* of discovery is neither unique in the laboratory nor in the logic of learning in general. In his analysis the logic of discovery is circumscribed and adumbrated by the logic of the psychology of learning. If such is the case, it seems that whatever the *special* value of discovery, this is not to be articulated in the logic of learning. As such, the logic of learning can tell us little about what we understand by discovery, especially in Hanson's sense when he suggests that discovery is what science is all about. Indeed discovery would seem to be least distinguished by virtue of its 'logic'.

Consider this. The biology student who conducts laboratory replications of celebrated experiments, though he initially experiences the anomalies for himself, and ultimately grasps from their study the overall patterns of relations by which they are governed, is a far cry from being himself a discoverer. Though he may exhibit all the developments prescribed in Hanson's model, these are not the conditions of discovery, though they may be a fortuitous circumstance associated with it. In the same light stands the tardy co-discoverer; the researcher who is not apprised of the value of his work; the 'forerunners' of great discoveries; as well as those who come to prominence for the wrong reasons. Clearly, these cases point to conditions of discovery that are unavailable to an inquiry limited to the logic of discovery. We shall return to this at greater length later.

Hanson showed some recognition of these situations in his last paper, 'An anatomy of discovery'.[8] Rather than extend and refine the model which was outlined in his earlier papers, Hanson embarked on a taxonomy of discoveries in terms of grammatical categories: discovering an X (e.g. a comet), discovering X (e.g. oxygen), discovering *that*

X (e.g. electrons disport themselves in an indulatory manner). Furthermore, he sought to classify the processes by which the anomaly becomes defined with a distinction between 'trip over' or accidental discoveries, and 'back into' and 'puzzle over' discoveries. However, the paper did not result so much in a further analysis of the logic of discovery as a generous stock taking of the complexities which his earlier writings had perhaps made appear overly simple. Judging from the grammatical preoccupation in 'An anatomy of discovery' Hanson appeared to be revising his linguistically oriented approach to the problem – one which reflected the thrust of Wittgensteinian thinking as opposed to gestalt models. After his accidental death, it appeared no one would come to grips with the problems he had identified in his last article – at least until the investigations of Richard J. Blackwell, a Notre Dame philosopher whose *Discovery in the Physical Sciences*[9] appeared in 1969.

BLACKWELL AND THE EPISTEMOLOGICAL APPROACH

Blackwell follows Hanson's lead fairly directly. His justification for the philosophical investigation of discovery is lifted directly from Hanson's rejection of Popper: the philosophy of science cannot confine itself to the study of the finished products in scientific explanation, nor the hypothetico-deductive model which entirely overlooks the processes of reasoning by which ideas are originally generated. Blackwell turns to the context of discovery itself, i.e. the 'investigation of what scientists have actually done in formulating new theories'.[10]

Based on an analysis of Descartes' and Newton's attempts to formulate the law of inertia, Blackwell specifies six components of the act of discovery:[11]

   i.   the selection of relevant circumstances
   ii. the discrete specification of relationships (Y if and only if X)
  iii. the idealization of concepts (e.g. point mass, instantaneous velocity, free falling bodies)
  iv. the integration of the hypothesis into a larger scientific framework or unit
   v. the precipitation of reasoning in accord with epistemological expectations
  vi. the reorganization of our understanding of the matter at hand

Blackwell then introduces a typology of discoveries, again based largely on Hanson. 'Discovering that X is the case' versus 'discovering why' are correlated with the explanandum and the explanans of the argument, and are paralleled by the further distinction between empirical and theoretical discovery. *Accidental* discovery is treated as a special case of 'discovering that', where the conceptual readiness for a puzzle solution is combined with the fortuitous circumstances of its realization.

The core of Blackwell's position emerges in his description of four possible levels for the explanation of discovery. These are: the logical, the historical, the psychological, and the epistemological. The four positions are represented by Hanson, Kuhn, Koestler, and Blackwell respectively. As we shall see, the epistemological framework which Blackwell chooses tends to obfuscate the issue of 'what scientists have *actually* done' in favour of describing the general grounds which, according to Blackwell, make discovery possible.

Regarding Hanson's attempt to specify discovery in terms of a process of retroduction, Blackwell notes that there is no *one* simple logic that exhausts the process, but probably many.[12] Blackwell consequently bypasses the inductive versus deductive question in favour of concentrating on what he calls the 'content-centered inferences' which 'play an essential role in the transition from data to the formulation of explanatory hypotheses'.[13] This is the epistemological approach. 'By this is meant an examination of the processes of discovery from the point of view of the meaning-content of the knowledge involved . . . and the factors involved in the transitions between these levels of meaning'.[14] 'The method needed in such an analysis might be called the method of conceptual gestaltism'.[15]

In spite of this promise of a gestalt methodology, which would confine itself to the real processes of reasoning employed by the researchers, Blackwell adopts a metaphysical position. What his approach produces in the end is an articulation of the widest parameters in terms of which the problem of discovery is seen to be given. For Blackwell, the cartesian dualism is primary; the basic structures for him are the human mind, physical reality, and their inherent compatibility. The physical sciences have been possible only because 'the human mind and physical nature are somehow appropriate for one another'.[16] Hence, the fundamental mind–nature confrontation is outlined as a process of the 'adaptation' of the mind to the perceived structures of nature. These adaptations are either the *elaboration* of existing structures (i.e. discovering X; normal science; empirical discovery); or the *transformation* of structures (i.e. discovering that X; revolutionary science; theoretical discovery).

In either case, the adaptation which is elaborated is conjectural; this is a direct consequence of Blackwell's formulation of the problem in the widest possible terms. While his analysis of Hanson correctly suggests that some number of logics may be attendant on discovery, so too any number of hypothetical structures may emerge as the 'meaning content' is spawned 'from the common meeting of mind and nature';[17] but not all of these will be valued as new discoveries. Some will simply be common sense inferences; some will be erroneous doctrines. Blackwell has traded his ideal of studying the actual processes of reasoning for a hypothetical model which, by stating the problem in its widest epistemological foundations, reveals no necessary accord with the actual experiences of historical discoveries.

As was noted earlier, Hanson's model of the process of discovery would be more correctly understood as a logic or model of learning. The same can be said of Blackwell's work. What he articulates is a methodology of learning couched in cartesian terms. Consider the characteristics of the *elaboration of structures:*[18]

    i.   curiosity
    ii. the sorting out process
    iii. interrelation of recognized portions of the structure
    iv. integration with other knowledge
    v. transcending actual experiences

These points might well appear in a psychiatric treatise on the growth of the socio-sexual maturity of the individual: the *curiosity* about the genitalia and the changes at puberty, the *sorting out process* in dating, the *interrelation* or 'integration' of the 'components of the structures' in heterosexual 'explorations', i.e. sexual relations, *the integration of knowledge* of sexuality with knowledge about reproduction, and finally *the transcendence* of actual erotic experience in the emergence of generalized affection and familial commitment. Or they might adequately describe the process whereby an aircraft mechanic traces the cause of an air crash from a broken bolt to a basic design flaw generated by hasty production and strong market pressures.

Consider also the elements of the transformation of structures:[19]

    i.   idealization of factual structures through conceptualization
    ii. creative postulation
    iii. substitution through analogy

This model is equally forceful in highlighting the transformations involved in the theoretical discoveries in science, as the gestalt switches experienced in religious conversion, political enlightenment and psychological catharsis. For example, a woman realizes that the hardships of her role as mother and housewife constitute a set of regular and unavoidable limitations on her life style. She postulates that these are essential features of the conjugal relationship in sexist society, and, substituting the model of the class struggle for the tensions of male–female experiences, her consciousness is revolutionized, and everything falls into an 'appropriate' political perspective. These fundamental revolutions in outlook are far from singular to science. Yet Blackwell is correct in designating them, along with the model of elaboration, to key roles in the mind–nature confrontation. However, whatever success is achieved at the general epistemological level by evidencing the foundations of our experience in terms of the pattern of interaction between mind and nature, little light is shed on the ways in which particular patterns of reasoning gained reputation and recognition as *scientific* discoveries.

General model versus specific cases: the use of ad hoc accounts

Though Blackwell's model covers every case of discovery (as well as learning to make love, conversion to a faith, solving a riddle, etc.) *in general*, in each particular case he is unable to show why an individual reasoned one way as opposed to another. Take the case of Descartes on inertia: 'That the selections by Descartes and Newton are different helps emphasize the need for some selection. This need does not explain of course why Newton defined momentum precisely as he did.'[20] However, to shore up the inability to specify the particular determinations of an outcome in his approach, Blackwell glosses the actual processes of reasoning with the cry of 'genius', 'insight', 'mystery', and 'luck'. 'There are major difficulties to be faced in any attempt to deal with discovery. Primary among them is the fact that the operations of genius and gifted insight have an air of uniqueness about them which is especially resistant to analysis.'[21] With reference to the failure of Descartes and the success of Newton in describing the laws of inertia, Blackwell notes that some initial selection of factors had to be made by both – with two different outcomes. Just how Newton selected those factors which proved relevant is a 'question of special insight and creative genius'.[22]

In a similar vein, in concluding his opening chapter, Blackwell alludes to 'the advance of science' as 'a mysterious process'.[23] Having attributed the core of real historical discoveries to mystery, it comes as no surprise to find this attribute at the heart of Blackwell's epistemology: 'In the act of discovery we should see concretely realized the interplay of mind and nature and should be able to see in at least one instance the mysterious harmony which exists between them.'[24] And lastly, when no case can be made for genius, insight and mystery, Blackwell resorts to the following: 'Certainly, it cannot be denied that

Lady Luck has occasionally smiled upon the men of science by providing situations that may otherwise have been overlooked or never deliberately contrived.'[25]

One is led to wonder whether it is the frown of the Lady that keeps all the 'average' scientists humble, by obscuring their path to greatness with *poor* luck. Hanson criticized the behavioural scientists for their purported distortion of the issues by 'charging it all to mystery words like "genius", "hunch", "insight", and "intuition" '.[26] However, Hanson, like Blackwell, was not himself entirely innocent of this charge. In his initial statement of 'The logic of discovery' in 1958,[27] he concluded his argument by berating both the pure inductionists and the pure deductionists, noting that the latter played down the role of 'logic and reason' in discovery,[28] whilst the former 'make it sound as though genius and insight have nothing to do with discovery'.[29] His discussion of the inductionists attributes a distinct role to genius in the production of *notable* discoveries.

Positing the source of discoveries in such 'imponderables' as genius or insight only obfuscates the model which laid such stress on anomaly, retroduction, and gestalt switch. Yet, in Hanson's estimation, cases of original, great discoveries were conditioned by genius. What an ironic position we have here – in every particular important case, the determining condition is the very thing which has been hounded from the general case. Additionally, if the only evidence of genius is the very discovery which it is used to explain, Hanson's account is not merely inconsistent, but the explanation is tautological. Furthermore, the account is empirically inaccurate, for who would want to claim that Descartes was not a genius, in spite of his failure?

In sum, our examination of Blackwell reveals that he is as guilty of Hanson's charge as is Hanson himself. And surely no outcome is more inevitable given the metaphysical backdrop for the study of actual, practical scientific reasoning. Given the direction of the inquiry toward the specificity, contingency and detail of what scientists have actually done (a direction chosen on the disillusionment with the logical empiricist position), any answer which relied on models characterized by an all-embracing inquiry into the foundations of knowledge and a theory of mind could not but fail to meet the requirements of specific historical processes.

KUHN'S MODEL OF SCIENTIFIC DISCOVERY AND CHANGE IN SCIENCE

Kuhn's work has provided an enormous impetus to reflection about the problems of change in science. Among philosophers his concept of the incommensurability of scientific theories and paradigms, and the theory-ladenness of observational terms, resurrected the relativism associated with Mannheim's sociology of knowledge. Both Popper and the neo-Popperians joined in the fray in the cause of Truth and Objectivity. Among sociologists his work focused attention on the substantive issues of theory change in science, and the role of social cleavages, science education, generational factors, and research networks in such changes. He has become as important in the sociology of science as Merton, and has directed sociological attention to the substantively scientific factors in the behaviour of scientists. The remarks I wish to offer are made from a sociological orientation. We must separate the issues which are being addressed in this treatment of Kuhn from those epistemological questions that are typically entertained in such discussions.

In this section we shall deal expressly with the model of discovery which he offers. However, it should also be noted that even in the sociological aspects of his work, there are two separate things going on: an account of how discoveries are made by individuals, and an account of how communities resist and/or accept such discoveries. The first question deals with a model of scientific discovery, the second with a model of social change. Our inquiry concerns the first question almost exclusively.

Paradigms and conceptual revolution

Central to Kuhn's model is the 'paradigmatic' structure of normal science. Normal science is described as the activity with which the great proportion of all scientific work is associated. In general it refers to the conceptual organization – explicit and implicit – in theories, and so includes their specified laws, their metaphysical foundations and world view, as well as the methodological strategies tied to them. In sum, the paradigm amounts to something of a comprehensive, scientific 'dogma'[30] to which the scientist is committed. According to Kuhn, discoveries occur when the conceptual suit of a paradigm no longer fits the body of knowledge, and unthreads in the form of anomalies.[31]

However, not everyone who is confronted with the unexpected observation realizes its value for reorganizing the theory. Such observations may not even register consciously; and if they do, they may be consigned to artifact or error. This was certainly the case in the card experiments which Kuhn refers to[32] in which the mismatch of colours and symbols went undetected in short exposure trials. An analogous situation was the repeated sightings of the planet Uranus on some seventeen separate occasions between 1690 and 1781 before it was realized that it was in fact a new planet. Consequently, Kuhn identifies two necessary conditions for the perception of anomalies. One of the 'normal requisites for the beginning of an episode of discovery . . . is the individual skill, wit, or genius to recognize that something has gone wrong in ways that may prove consequential'.[33] And secondly, 'anomalies do not emerge from the normal course of scientific research until both instruments and concepts have developed sufficiently to make their emergence likely and to make the anomaly which results recognizable as a violation of expectation'.[34] The second phase of the process is usually a temporally protracted attempt to ascribe law-like properties to the anomaly through repeated observation, experimentation and thinking.[35] Finally the new discovery comes to fruition in the emergence of a radically new conceptual model which explains the new anomalies and which, because it was virtually 'pre-paradigmatic', could not have been predicted beforehand.[36]

Since anomaly is the most important concept in the expedition of discoveries, we shall examine this idea in some detail.

Anomaly or novelty?

After the original recognition of the anomaly, Kuhn suggests that the process of discovery

continues with a more or less extended exploration of the area of anomaly. And it closes only when the paradigm theory has become adjusted so that the anomalous has become the expected. Assimilating a new sort of fact demands a more than additive adjustment of theory, and until that adjustment is completed – until the scientist has learned to see nature in a new way – the new fact is not quite a scientific fact at all.[37]

This passage is *the most elaborate* explanation of anomalies that one finds in Kuhn's theory. Although Kuhn does provide numerous *examples* of anomaly in his discussion of particular discoveries, I take the shortcomings of these examples to be that one grasps their relevance *by virtue of hindsight*, without understanding how they became anomalous in the first place. For example, the fact that Herschel, whom we partly credit with the discovery of Uranus, had previously observed 'either a nebulous star or perhaps a comet'[38] does not recover how his exploration of the northern heavens took on the value or identity he assigns to it. In retrospect, the retroduction from anomaly to discovery seems a natural development or imperative. However, if we bracket the assistance of hindsight it seems plausible that on the first sighting, indeed in those first few seconds, Herschel might have had pause for reflection. This was an unexpected sighting or novelty; no more. It could have been 'seen but not noticed',[39] overlooked or ascribed to error. An adjustment for artifactual light might have been made. Also, a comparison might have been made with a star in the same region, to confirm the size of the object. In such a situation a friend or assistant might have been called over to confirm the observation. What emerges from this speculation is a complex set of alternative developments each of which could have terminated the sequence of inference prematurely. For example, the observer might have failed to observe the phenomenon and it might have been ascribed to error, or if seen, it might have been perceived as a *factual* novelty or curiosity. Or it might have been an anomaly or puzzle which simply fizzled out, or it might have led to a qualification of an existing theory, or to something completely different. Kuhn fails to examine all but the last of these possibilities.

On the other hand, the practised eye might have required no adjustment or reformulation, given the perceived reliability of the telescope, the competence of the observer, and his knowledge about the categories of things that one could possibly observe under such circumstances. In such a case the anomalousness would be immediate and apodictic. Nonetheless the conditions for the identification of the observation as an anomaly would remain to be articulated through an examination of the pre-reflective background into which the novelty emerged. As noted earlier, Kuhn glosses this transition with an oblique reference to the 'individual skill, wit or genius' of the researcher. This transition from novelty to anomaly requires some discussion.

'Anomaly appears only against the background provided by the paradigm'.[40] This statement can be re-iterated to centre the mutual dependence of paradigm and anomaly of interest here: anomaly makes present the paradigm through which it is seen. In other words, Kuhn's idea of anomaly is not just the observation of a novel fact, but amounts to something like a suspicion or first doubt regarding the adequacy of the conceptual order in which the novelty ought to have been inscribed. Hanson's concept had this same meaning. In common usage, anomaly is simply an unexpected observation or irregularity; however, for Kuhn and Hanson, when an irregularity is assigned anomalous status, this has more than the common implication. It is already a tacit renunciation of the prevailing model. This would suggest that anomalies themselves are discovered. However, if we were to admit that this must happen, then we would displace the question of the conditions for discovery from the object discovered to the discovery of the anomaly! These conditions, however, must be the same, for in Kuhn's account, the act by which an observation is figured as an anomaly is the same act by which the discovery is initially grasped. In other words, it appears that nothing is really *explained* about discovery by claiming as does Kuhn that anomalies are necessary conditions for discovery, for given Kuhn's sense of anomaly this amounts to saying that a discovery must be discovered in order for it to be a discovery in the first place, and this is plainly circular.

Nonetheless, the model of discovery which Kuhn advances is quite complex. He rejects the notion that discovery is a 'unitary event, one which, like seeing something, happens to an individual at a specifiable time and place'.[41] Indeed, several persons may be involved over an extended period of time.[42]

According to Kuhn, as we have seen, the discovery begins with the definition of paradigmatic anomalies, the elaboration of their value, and the reorganization of the theories for their accommodation. As such, Kuhn leads us to treat the relationship between anomalies and revolutionary innovations as causal or determinative in character. If this is the case, one might say that the detection of anomalies is a necessary but not a sufficient condition for discovery. It is necessary in that without their detection, a new theory would never be proposed, given the apparently satisfactory operation of the existing ideas. However, anomalies are not sufficient to initiate a revolution in science. Here Kuhn introduces laterally a different condition: it is not enough that a researcher take note of anomalies (in the common sense), but he must be reflective of his own cognizance of them as anomalies: 'Apparently, to discover something one must also be aware of the discovery and know as well what it is that one has discovered.'[43] Kuhn introduces this remark after he notes that Stephen Hales had successfully isolated relatively pure samples of oxygen some forty years before either Scheele, Bayen, Priestley or Lavoisier had done the same thing. However, Hales neither classified this gas, nor treated its isolation as central to chemistry. He was not especially aware that he had found anything earth shaking, even though he had gone through all the motions followed decades later by Priestley, and recorded the novelty of the substance. In other words, the novelty of the gas was not an anomaly for Hales.

This proposition introduces problems for Kuhn's model. The second condition for discovery, that the 'anomalies do not emerge from the normal course of scientific research until both instruments and concepts have developed sufficiently to make their emergence likely', is *true by definition* because an anomaly is a novelty which calls into question the conceptual basis by which it is produced. In other words, if there is no background developed to give the novelty relief as an anomaly, none will be seen. This is intimated in Kuhn's suggestion that one must 'be aware of the discovery',[44] as well as know 'what it is that one has discovered'. Surely, however, only one point is made here. Kuhn cannot mean that Hales inadvertently and unconsciously was to be found sleep walking in his laboratory, unaccountably isolating new substances. What is meant by awareness is surely the second thing which he mentions: one must know 'what it is that one has discovered' – one must be aware of the identity of what one has found. An identification of the substance by Hales was, however, *not* absent. He identified it as another gas which composed the atmosphere. Nonetheless, Kuhn does not recognize this work as a discovery for, from the benefit of hindsight, the real conceptual advance was made in Lavoisier's work on combustion. This introduces another point. At the outset Kuhn circumscribed the domain of discovery to include *only* conceptual advances (which he describes as 'my exclusive concern') and to exclude enumerative identifications of factual novelties. He continues: 'there is another sort and one which presents very few of the same problems. Into this class of discoveries fall the neutrino, radio waves, and the elements, which filled empty places in the periodic table'.[45] Hence, Kuhn's model of discovery implies at the outset the sort of phenomena that he will consider: 'inventions of theory' which reorganize factual domains. In the case of the discovery of oxygen, the real question for Kuhn is not who isolated the element *per se,* but rather, for whom was the appearance of such an element a source of major conceptual reorganization? Clearly the answer points to Lavoisier for whom oxygen became central to the theory of combustion and to the overthrow of the phlogiston theory of burning.

Hence by implying that Hales had not experienced an anomaly, Kuhn by definition means that Hales was not aware of the paradigmatic consequences of his observations. Specifically, Hales was not aware that the new gas had special combining properties which explained combustion. This realization constituted the chemical revolution initiated later by Lavoisier.

The supposition which emerges from the above is that what Kuhn portrays as ostensive elements of a relationship, and conditions for discovery, are in fact aspects of his definition of discovery. However, this possibility is precluded in Kuhn's account by virtue of his use of historical examples.

Thematic aspects of the examples

As indicated earlier, the examples in Kuhn's account bear the brunt of the explanation. By the use of examples (for instance the discovery of oxygen), Kuhn evades the formal identification of the properties of discovery by assigning to discoveries in general the properties of the particular examples he chooses to review. In the account of the discovery of oxygen, Hales, Bayen, and Scheele are excluded from the discussion at the outset;[46] Lavoisier and Priestley dominate the question, because it is from their interaction that the theory of combustion later emerges. In the theory of combustion, oxygen is the key element. For all the earlier researchers, oxygen was, by contrast, just a new element among others. It is interesting in this respect that Kuhn switches his discussion of the discovery of Uranus in his 1962 article in *Science* for an outline of the history of research on Leyden jars and electricity in *The Structure of Scientific Revolutions*; presumably, the observation of Uranus was not actually an anomaly. Hence Kuhn's examples disguise a technical use of the word 'discovery', for they only exhibit discoveries which were also conceptual revolutions.

Furthermore, Kuhn's use of historical examples has no small rhetorical function, inasmuch as what are identified analytically as *necessary* conditions figure in the historical account as both necessary and *sufficient*. After all, their adequacy seems to be an indisputable fact. Given that someone has made a great discovery, he must have had a sufficient degree of wit, skill or genius, the conditions of the paradigm must have been sufficiently developed, and the anomalies must have been sufficient to lead to a re-evaluation and, finally, a revolution of scientific knowledge. That is, given the historical conclusion or outcome, all the premises must be true; after all, we are already living with the consequences. This sort of argument, however, is logically fallacious. 'It is a logical error to infer the truth of the premises from the truth of the conclusions.'[47]

Consider a counter example:

i.	if a man is bitten by a rattlesnake	(antecedent)
	then he'll die	(consequent)
ii.	a man dies	(an observation)
iii.	therefore he has been bitten by a rattlesnake	(conclusion)

The conclusion is erroneous. It cannot be deduced from the observation in (ii), for any number of antecedents could suffice for the outcome. Yet it is this fallacy which is obscured by the historical examples. The fact of the outcome, the discoveries, the consequent of the argument, does not demonstrate the relevance of the antecedents, though if we were in a situation where the antecedents were observed (i.e. anomalies), we would predict the consequent (discovery).[48]

KOESTLER ON THE ACT OF CREATION: BISOCIATION OF IDEAS

The circularity of the gestalt accounts which we have noted in Hanson and Kuhn is avoided in Arthur Koestler's theory. For Koestler, discovery and other acts of creation *may* occur as gestalt shifts, but 'gestalt' *per se* has no explanatory importance. Gestalt shifts are the result of thought processes which are not always fully conscious, but are sub- or pre-conscious.

Koestler argues that the thought processes which underlie artistic creations and comic inspirations, as well as scientific discoveries, are *structurally* identical. They result from the synthesis of a single idea with two apparently inconsistent contexts. Koestler coined the word 'bisociation' to describe this. These syntheses cause at first a mild aggravation, distraction or sense of displacement that is 'released' by an emotional 'flood out' whose quality is dependent on the definition of the situation. For example, jokes produce a sense of juxtaposition which is laughed off. Discoveries consist in a synthesis of ideas that is characterized by a mix of elation and catharsis. And literary creations consist of images, representations, and analogies which lend a sense of poignancy to the commonplace and elicit an aesthetic or cathartic experience. In each case the bisociation creates an element of tension which is released or expressed in the pertinent emotional form. Koestler's case begins with an extended analysis of the bisociative structure of everyday humour and proceeds by analogy to science. The following 'specimen' is drawn from Koestler.[49]

Two women meet while shopping at the supermarket in the Bronx. One looks cheerful, the other depressed. The cheerful one inquires:
    'What's eating you?'
    'Nothing's eating me.'
    'Death in the family?'
    'No, God forbid!'
    'Worried about money?'
    'No . . . nothing like that.'
    'Trouble with the kids?'
    'Well, if you must know, it's my little Johnny.'
    'What's wrong with him, then?'
    'Nothing is wrong. His teacher says he must see a psychiatrist.'
    Pause. 'Well, well. What's wrong with seeing a psychiatrist?'
    'Nothing is wrong. The psychiatrist said he's got an Oedipus complex.'
    Pause. 'Well, well. Oedipus or Shmoedipus, I wouldn't worry so long as he's a good boy and loves his mama.'

In this example Koestler argues that the humour consists in 'bisociating' the single phonetic utterance about how a son should love his mother in two inconsistent contexts: the normative relationship of mother–son affection, and the context where such affection is pathological. Koestler shows that the same sort of juxtaposition of inconsistencies is bisociated in other types of humour: for example, in pun and witticism, in impersonation, in satire and caricature, in role inversions, and even in tickling.[50]

Scientific discoveries are *structurally* the same. Discoveries according to Koestler are syntheses of commonplace events with theoretically important contexts. The leading illustration of this point is Archimedes. The change in the water level produced by stepping into the gymnasium bath was certainly something with which any experienced bather must have been familiar. However, Archimedes bisociated his observation of the change in water level with the question of the proportion of gold in Hiero's crown. In other words, the commonplace displacement of bathwater became thematically important for the solution of a problem in another context. However, the inadvertent solution to the problem in the context of bathing constituted a surprise, a disruption of the mundane course of bathing. This accounts for the classical reports of Archimedes' distraction and exuberance.

Koestler further documents the bisociative nature of great discoveries by examining the circumstances under which the work of Gutenberg, Kepler, and Darwin proceeded. In each case, he shows that the discovery was made through an apparently inadvertent synthesis of facts. Thus Gutenberg, after witnessing the awesome power of the wine press, displaced its use to that of the carved wooden seal used in transferring images. This was, quoting Gutenberg, 'a simple substitution which is a ray of light': the *printing* press.[51]

Darwin was much struck by the infinite number of species and varieties of living forms and how they seemed to vary from one another by degree as though descended from a common source. Yet he was unable to account for the principle which underlay the assortment of creatures. He recounts that it was the reading of Malthus and the gloomy image of the constant struggle for existence which made all the parts of speciation fall into place. In other words, he bisociated the account of human population dynamics to the problem of animal and plant variability, thereby laying the foundation for *The Origin of Species*.[52]

So too Kepler resolved his failure to correct the elliptical orbits of Mars, which he initially believed on metaphysical grounds to be a perfect circle, by drawing the unorthodox conclusion that astronomical bodies obeyed *physical laws* and hence that the orbits of the planets were governed by physical forces. This bisociation of physics and astronomy laid the groundwork for the seventeenth-century revolution in physics.[53]

Gestalt and the unconscious

Like Max Wertheimer, Koestler pays serious attention to the character of scientific insights – namely that they often occur as dramatic turns or 'flashes' of awareness, that is, as gestalt switches. Indeed, Koestler prefaces his discussion of Archimedes with the exemplary cases of 'gestalt shifts' – Köhler's report on the insight thinking of his celebrated chimps, Neuva and Sultan. However, Koestler argues that these occur as a consequence of an unconscious or subsidiary preoccupation with the problem. On this point he follows Poincaré's well-known account of his discovery of Fuchsian functions. This deserves to be examined at length.[54]

For fifteen days I strove to prove that there could not be any functions like those I have since called Fuchsian functions. I was then very ignorant; every day I seated myself at my work table, stayed an hour or two, tried a great number of combinations, and reached no results. One evening, contrary to my custom, I drank black coffee and could not sleep. Ideas rose in crowds; I felt them collide until pairs interlocked, so to speak, making a stable combination. By the next morning I had established the existence of a class of Fuchsian functions, those which come from the hypergeometric series; I had only to write out the results, which took but a few hours. Then I wanted to represent these functions by the quotient of two series; this idea was perfectly conscious and deliberate, the analogy with elliptic functions guided me. I asked myself what properties these series must have if they existed, and I succeeded without difficulty in forming the series I have called theta-Fuchsian.

Just at this time I left Caen where I was then living to go on a geologic excursion under the auspices of the school of mines. The changes of travel made me forget my mathematical work. Having reached Coutances, we entered an omnibus to go some place or other. At the moment when I put my foot on the step the idea came to me, without anything in my former thoughts seeming to have paved the way for it, that the transformations I had used to define the Fuchsian functions were identical with those of non-Euclidean geometry. I did not verify the idea; I should not have had time, as, upon taking my seat in the omnibus, I went on with a conversation already commenced, but I felt a perfect certainty . . .

Most striking at first is this appearance of sudden illumination, a manifest sign of long, unconscious prior work. The role of this unconscious work in mathematical invention appears to me incontestable.

The unconsciousness in this account and in Koestler's is decidedly *not Freudian*. That is, no hard and fast line is drawn between the structure of conscious awareness and the subconscious mind. On this point, Koestler notes that *'awareness is a matter of degrees'*.[55] Similarly, Einstein suggested that full consciousness of a topic is the limiting case[56] – indicating that most thought sways back and forth between attentiveness and reverie. This notion is also central to Michael Polanyi's discussion of the tacit dimension of knowledge, the knowledge which reposes not in full consciousness but in our 'subsidiary awareness'.[57] Polanyi implies that discoveries are rooted not in the methodological fetishism of the philosophers of science, but in the often inexplicable commitment of an individual or community to a paradigm which may at times be at variance with the facts.[58]

All this bears directly on the idea of gestalt switch. Clearly, Poincaré's experience and that attributed to Archimedes are not inconsistent with the revolutions in consciousness seen in Köhler's apes. However, as noted in the previous section, the identification of such events by merely titling them does not account for them. The argument from the 'sub-' or the 'pre-' or the 'subsidiary' consciousness does. In the textbook cases of visual gestalts, for example the goblet which reverts into a symmetrical profile of two human faces, we are led to believe that a gestalt consists in the ordering of elements into a figure–ground relationship, and gestalt shifts consist in the displacement of our focus of attention when we reverse the figure–ground order. The implication is that the mind can consider only the one image *or* the other, and hence that consciousness is monolithic. This implication is borne out by the assumptions of Kuhn, Hanson, and Feyerabend regarding the gestalt switches in 'paradigm changes'. It is held that adherence to one paradigm excludes perception of other paradigm views – as though we could only bring a single paradigm to mind at one time.

Koestler, however, suggests that though most will not be the direct topic of awareness, several things can be 'on one's mind' simultaneously. Indeed, Koestler holds that awareness is a matter of degree largely because much of our activity is conducted from routine and habit. We do things – like boarding a bus – 'automatically', or *without* conscious reflection and deliberate execution.[59]

Hence the version of gestalt that Koestler illustrates is more akin to Schutz's idea of multiple realities[60] which, as interpreted by Goffman,[61] focuses on the moment to moment changes in one's awareness and changes in one's presence of mind, from the private sphere to the world of one's dreams and recollections. In this gestalt shift, the appearance of an idea may be *temporally* inadvertent but *thematically* it is altogether in order with something that we have 'on our minds' – though not necessarily right at the moment when 'it comes to us'.[62]

Consequently, the switch in Archimedes' attention from what was probably the routine and habit of bathing to the problem of Hiero's crown is provided for by the picture of consciousness which Koestler depicts – a consciousness of various degrees of awareness, which in the stream of experience evokes many contexts and levels of reality and various topics simultaneously.

The problem of unconscious judgement

According to Koestler, the unconscious is the ultimate 'matchmaker' in the arrangement of bisociative synthesis, for the unconscious, unlike the rationally reflective consciousness, is not hampered by routine, logic, or proportion, but proceeds by unconstrained forms of ideation, the substitution of mental imagery for strict verbal formulae, symbolization, mixed imagery, and unbridled analogy.[63]

However, precisely because the unconscious is not guided by logical rationality, many of the episodes of spontaneous illumination may turn out to be erroneous. For example, Pasteur reasoned correctly *by analogy* that the resistance of a batch of chickens to cholera was similar to the resistance of victims of cow pox to small pox, i.e. they had been 'vaccinated' (from *vacca*, cow).[64] Likewise, Semmelweis reasoned correctly that the symptoms of infection produced by a lesion during an autopsy were *analogous* to the symptoms of child-bed fever among post-partum women.[65] However, Leverrier's prediction of the planet Vulcan to explain the precessions of Mercury at perihelion, which was based on the model which had proven successful in predicting the existence of the unobserved Neptune from the perturbations of Uranus, met with no success.[66] Though the analogy was a flash of inspiration, it proved to be of no avail.

Poincaré sidestepped this problem in his famous account of mathematical discoveries.[67] It was his suggestion that the subconscious mind was *superior* to the conscious ego in determining the worth of an idea. He speculated on the basis of his own experience that the subconscious ego was directed by the mathematician's sensibilities of aesthetic form and proportion, and that of all the possible syntheses which could be determined, only those which were aesthetically pleasing would be discriminated for presentation to the conscious mind. According to Poincaré, their aesthetic form constituted 'the presentiment of a law'. This subconscious sensibility was identified as the root of Poincaré's own successes. However, in light of Leverrier's case, and in light of Poincaré's own failure to appreciate the far reaching consequences of the asymmetry of Maxwell's equations (i.e. his near miss of special relativity theory), Poincaré's claims overstate the case for the infallibility of the subconscious mind.[68]

The argument for the occurrence of discovery based on the action of the subconscious and its eruption in gestalt switches can provide no more than a necessary condition for the explanation of discovery. That is, discoveries belong to the set of creations with gestalt shifts and subconscious inferences; however, as Koestler himself notes, this class also includes comic inspiration and artistic inventions. As we noted with Hanson and Kuhn, scientific discoveries are clearly *not* distinguished on the basis of the characteristics of the thinking which gives rise to them. Nor is it established that all discoveries are based on gestalt switches and subconscious ruminations. Urey reasoned logically and deductively that a *heavy* atom of hydrogen was implied by the structure of the periodical table;[69] his discovery of such an element seems to have been the very antithesis of Archimedian inspiration. Consequently, it seems difficult to hold in principle that gestalt and the subconscious are *either sufficient or necessary* conditions for scientific discovery, for not all gestalt switches are scientific and not all discoveries result from gestalt switches.

Perhaps the problem here derives from looking at all things called 'discovery' as though they were essentially similar. It may be that the only thing they have in common is what they are called, just as tools in a toolbox are massively dissimilar except for being called 'tools'. Consequently, the idea of a nomic relationship which is uniform for all discoveries is probably erroneous, at least within a naturalistic perspective.

We have examined several writers with ideas about a model of discovery. Some of our objections have appeared in each work, others not. What I propose to do is to itemize each of the problems described in the above accounts and, by examining them together, provide an overall assessment of the costs involved in limiting ourselves to the models proposed in these works.

3
A synthetic assessment of the psychological accounts

The models of discovery explored in the last chapter have certain common features. All appear to be motivated in some way by the Reichenbach distinction; specifically, they are conceived as responses to the vacuum created by the traditional preoccupation with the context of justification. All draw attention to the logic-in-use of scientists in their research. And all attempt to explicate this *in situ* activity with reference to historical examples. Consequently, each author effectively outlines a model of discovery, that is, an explanation of the circumstances under which discoveries have occurred.

Furthermore, though there are important differences in the exact formulation of the models, all these models nonetheless constitute the same *kind* of explanation, namely, a mentalistic explanation. Whether the controlling variable is called gestalt shift, the perception of anomaly, retroduction, ostension, skill, wit, genius, insight or luck, all these accounts offer explanations of the occurrence of discoveries by showing how, as a result of interaction with the environment, new ideas get into the researcher's head. Hence these accounts are all reductionist. That is, they reduce the problem of historical discoveries to the psychological level. This strategy, while it has costs, as we shall see in a moment, also has an interesting benefit: it suggests that every and *any* psychological model which accounts for how an individual appropriates a new idea could also constitute a model of scientific discovery. In other words, any advances in psychology, information processing theory or the cognitive sciences which broadened our understanding of the knowledge-acquisition process would simultaneously be advances in our understanding of the 'logic' of discovery as defined earlier.[1] Seen in this light, it is not really surprising that so little attention has been paid to the context of discovery compared to the context of justification, and that the contributions made to our knowledge of the latter far outshadow those made to the former. While it makes sense that there should be a consensus about what makes a sound theory, it really doesn't matter how diverse the kinds of thinking are which gave rise to it, nor does it matter whether these are even reported in the theory. After all, the scientists can employ any method whatsoever of arriving at an idea. These methods are as extensive as learning itself, and indeed are equivalent to learning in all its forms. The only thing which differentiates the logic of discovery from ordinary learning is the nature of the thing learned. Consequently, we begin to see that the attractiveness of the problem of the context of discovery probably is not to be found in the actual psychological processes by which it is subtended, but in the intrigue in the topic of discovery itself. Indeed, discovery seems to be distinguished least of all by its character as a type of learning or idea acquisition. What is most intriguing about it is its status as a *social* phenomenon. We shall deal with this later. Having examined the psychological models, we will now consider a series of liabilities which detract from the utility and relevance of this approach to the problem.

Reductionist models are insufficient in principle

Our examination of accounts of discovery offered by Hanson, Blackwell, Kuhn, Koestler, and others points up numerous minor deficiencies. However, there is a *major* deficiency at the very core of such approaches. In showing that all these accounts, in spite of their philosophical, historical or psychological commitments, are mentalistic, we find that they are *in principle* incapable of accounting for discovery. This deficiency is revealed when we consider the causal value of any of these conditions. It was pointed out that whether the crucial condition was retroduction, perception of anomaly or subconscious synthesis, this could constitute only a *necessary* condition of discovery, because discoveries belong to the class of inferences with gestalt/retroductive/bisociative characteristics. Yet this condition would not be sufficient, because all members of such a group are not necessarily also discoveries. As Hanson indicated, these groups include the schoolboy who experiences insight during his laboratory replications of classical experiments, as well as the original discoverer. Similarly, it was noted that Blackwell's two models of discovery apply as equally to such things as psychosexual development and political enlightenment, as to the 'elaboration' and 'transformation' of scientific knowledge. So too, Koestler's approach expressly acknowledged the relevance of his explanation to non-scientific fields, especially humour, art and literature. Consequently, these accounts fail to identify what is *unique* about discovery by focusing on what it has in common with non-scientific modes of thought. In other words, by focusing on what is necessary, from the point of view of the author in question, mentalistic approaches forfeit what is sufficient. The implication of this *formal* limitation is evidenced by the following *substantive* confusion.

Mentalistic models confuse learning with discovery

Reductionism in the natural sciences has had enormous benefits. However, in the social or behavioural sciences we tend to obfuscate the social significance of familiar phenomena by explaining them in terms of 'underlying' causes. Though this is not always the case, it is true with discovery and learning. As noted above, by examining the context of discovery under a mentalistic model, we tend to identify cognitive conditions or causes which characterize all occasions of learning, and in so doing shift the locus of the problem from the realm of scientific inference to general psychology. Presumably this consequence is avoided by the post hoc orientation: we typically begin an analysis with what are already discoveries, and hence, with our attention riveted to the case at hand, we are not distracted by the general grounds of the mentalistic approach. Nonetheless, when examined formally, this equation of learning and discovery is problematic. In fact, when viewed in historical perspective, the equivalence of learning and discovery is a *confusion*. From a social perspective, 'to *learn*' means something quite different from 'to *discover*'. Similarly to be a 'tardy' researcher who reports results long after they have been published elsewhere is to be quite different from the discoverer who first published such results. So too, the status of a 'forerunner' who, like Aristarchus, anticipates the writings of a later notable discoverer, in this case Copernicus, is quite different from that of the discoverer actually credited with the discovery. These different social statuses and the cultural meanings attached to them are what is obscured in the reductionist accounts. Plainly, an account of how Mendel learned the laws valid for *Pisum* would not be an explanation of the context of their discovery; and the mentalistic gymnastics which allowed the culprit to put together the Piltdown hoax would contribute little to the context of its discovery. These confusions obscure the social *statuses* attached to the events by giving primary consideration to the *mental correlates* which necessarily attend their production. Obviously, knowledge of this mental/cognitive activity would tell us little about why these events were discoveries, and how they occurred. That question requires a cognizance of the social status of discoveries versus learning, and this is something overlooked by the reductionistic strategy. So too is the following.

Mentalistic models apply to discoveries and errors indiscriminately

Not only are reductionistic or mentalistic explanations typical in non-scientific areas, and relevant to learning generally, but the same kind of gestalt switch that characterizes a success may apply equally to scientific failures and errors. Recall those intuitive flashes like Leverrier's unsuccessful prediction of Vulcan.[2] As we have seen, Leverrier had earlier predicted the existence of the planet Neptune based on the perturbations in the orbit of Uranus. He correctly attributed the unexpected ellipses in the orbit to the gravity of an undiscovered planet whose size and distance he correctly calculated before it was actually sighted by telescope. Confronted with similar anomalies in the orbit of Mercury, he posited analogically the existence of a small planet even closer to the sun – but none was found. This insight, no matter how aesthetically compelling and analogically sound, proved to be false.

We have a similar situation with the case of Poincaré. Poincaré, who attributed the source of mathematical discoveries to an infallible process of subconscious synthesis, appears to have been an exception to his own theory. Though keenly interested in the problems of relative motion and absolute space, and master of physics, astronomy and mathematics, he missed the breakthrough in relativity which was achieved by his younger contemporary, Albert Einstein.[3] Consequently we see that though the ideas of Leverrier and Poincaré were the products of some mentalistic process, they were also erroneous scientific ideas. Though they may have occurred through the same process, their social status was an enormously different social fact.

The causal adequacy of gestalt switch

We have spoken until now of the necessary-conditionship of mentalistic models. However, it appears at times as though certain uses of these causes are either tautological or merely descriptive. In its simplest expression this takes the form of a recommendation that discoveries occur as the result of gestalt switches. Since the switch is the thing which is occurring, it is circular to identify it as a cause of this. Though tautological statements have been patently obvious in the attribution of discovery to genius or luck, they can also be detected in the central explanatory concepts we have examined, where they frequently appear to disguise as natural processes what are in fact definitions or descriptions of discoveries. For example, consider Hanson's concept of 'retroduction'. The term refers to the apprehension of a relationship or phenomenon *retrospectively*. The researcher begins with an anomaly, as in Kuhn's theory, and then proceeds to reflect on what conditions would have to hold for the anomaly to have appeared of necessity. This involves the reinterpretation of a *prior* observation. The induction consists in determining the meaning of an event from *retrospect* – hence retroduction. However, like Max Wertheimer's concept of gestalt,[4] retroduction as such constitutes little in the way of explanation. To say that a discovery occurred through retroduction is to say that a discovery followed a period marked by the reconstruction of an anomaly. The discovery had not occurred at time t_1 but at t_2 it had. This is equivalent to saying that an anomaly was observed which later in time proved decisive. Though this might have some accuracy historically in showing *that* events occurred, it is not a principle of *how* such events occur. In other words it might be descriptive, but not explanative, because in Hanson's work, it does not explain how an anomaly becomes connected with a discovery. The only value of the title of the concept, 'retro-duction', is that it indicates that howsoever the thing happens, it takes time.

We have come across analogous problems in Kuhn's model. These centre around Kuhn's distinction between 'anomalies' and 'novelties'. It was observed that an anomaly constituted not just a *temporally* inadvertent observation or novelty, but that it constituted a *thematically* problematic phenomenon which motivated a course of research directed expressly at a possible explanation of the phenomenon. Hence the irradiation of barium plates was *anomalous* for Roentgen, but merely *novel* for certain earlier researchers.[5] Similarly, the properties of oxygen were *novel* for Hales, but were *anomalous* for Lavoisier.[6] In other words, when something has resulted from a novel observation, it is reconstructed retrospectively by Kuhn as an anomaly. Thus we see that where Kuhn represents in the concept of anomaly a relationship between anomaly and discovery, this in fact is part of the definition of discovery – for he defines anomalies as those novelties which motivate successful discovery. This problem is compounded further by another definitional peculiarity; for Kuhn, discoveries are *theoretical* achievements which entail *by definition* a reflective reconstruction of experience. Excluded by fiat are simple factual discoveries.[7] Hence the appearance of new elements would be simple *novelties* – which do not produce 'discoveries' – for discoveries are by definition theoretical breakthroughs and anomalies are those types of novelties which in the guise of explaining discoveries, only define them.

Factual discovery and necessary-conditionship

There is a further question that arises from our consideration of so called 'factual' discoveries. It could be argued that numerous discoveries have occurred merely by the enumeration of novel elements. I would suggest that in these cases the discoveries did not occur as a result of gestalt switches or conceptual reorganizations, and therefore that such mental processes are not only insufficient to explain discovery, but may also be *unnecessary*. For example, consider the discoveries which resulted from pointing the terrestial telescope at astronomical bodies and pointing microscopes at biological bodies. Galileo determined that the visual features of the moon were craters, mountains and valleys, that the brightness of the planets varied with their orbital placement, that there were spots on the surface of the sun, and that Jupiter had four satellites.[8] Similarly, following the Indian Wars and the extension of the railways throughout the western American plains, enormous fossil finds containing numerous unprecedented dinosaur specimens came to the attention of palaeontologists; these specimens were found strewn throughout the badlands and were identified without the necessity of gestalt shift, retroduction, etc.[9] In other words, they were discovered by simple ostension and enumeration.

By downplaying the role of gestalt shift in such cases, we are not claiming that the discovery involved no thought process whatsoever. On the contrary, what is being claimed is that the particular pattern of thinking described as gestalt shift was not required; this type of discovery was not apprehended by a stroke of intuition which forced its way into the imagination, reorganizing the conceptual and perceptual field. Yet this does not mean that diligent thinking and careful observation were uninvolved. However, it would be an error to assign such thinking a necessary-conditionship status for two reasons. First, thinking is probably a necessary condition for *any* social activity, and consequently there is nothing especially useful in singling it out as a condition of the activities we are examining. But secondly and more importantly, its very ubiquity suggests that it is a correlative of discovery action – and not a 'cause' of it. This observation broadens our earlier conclusion regarding gestalt shift. Just as gestalt shift is not a cause of discovery, neither is thinking *per se,* even the simple enumeration of factual novelties under consideration here. This thinking is part and parcel of the thing we are referring to as 'discovery', and not the 'cause' of it.

The post hoc status of mentalistic accounts

Since many of the above problems are only apparent when we suspend the benefit of hindsight, they typically are missed by those who are indebted to it for the force of their analyses. Usually, when we consider the problem of discovery we begin with bona fide historical specimens. By directing our attention exclusively to the scientist who has succeeded we risk a range of methodological biases. We tend to assume that there is something peculiar or extraordinary or decisive associated with this one, which explains why he succeeded where others failed. It rarely occurs to us that discoveries are made almost exclusively by those who are devoting their lives to that very end, and that particular discoveries appear when traditions motivate such research. These are excellent but all too ordinary predictors of discovery. As noted earlier these pithy thoughts seldom appear relevant in the usual approach to the problem which assumes a peculiarity in the successful researcher. Presumably this peculiarity, whether it be his eccentric personality, the serendipitous conditions of the realization, the uniqueness of the subconscious synthesis, or whatever, is regarded as the *source* of the discovery! In other words, by beginning from hindsight with an event which is of interest because of its singularity, we risk assigning its origins to whatever other singularity is associated with the event or the individual. Consequently, the post hoc status of the explanation is subject to certain potential errors. As noted above we often fix our attention on the thinking process itself – for it appears to be an unusual correlate of the event. Hence, it is tempting to treat that thinking which is the discovery as its cause. This leads us to conclude from hindsight that what may be necessary conditions are actually sufficient. This error was noted especially in Kuhn's case where historical examples lent to events in retrospect a sufficient-condition status, which analytically was only a necessary-condition status. All this of course can be glossed from the benefit of hindsight. For example, we know from hindsight that Hales did not initiate the chemical revolution which was based on the theory of combustion; this was done by Lavoisier. Consequently, he couldn't have had the skill, wit or genius to reason out the thematic relevance of oxygen to combustion; oxygen must have been no more than a novelty for him. But Lavoisier must have had the skill, wit or genius to grasp the anomalous character of oxygen. For was it not *he* who succeeded where others failed? Who can deny Lavoisier's achievement? Whatever the conditions for the discovery, were they not sufficient, inasmuch as the event transpired? Certainly if the event is lawful by nature, then if it has occurred, the conditions necessary and sufficient for its occurrence must have been present of necessity. So that if skill, wit or genius is identified as a condition of discovery, and the discovery has occurred (which is taken for granted when working historically or in retrospect), then the efficacy of skill, wit, or genius seems to be incontrovertible. Indeed, it appears as the decisive factor. However, as we have seen, the development of anomaly from novelty, or gestalt switch from anomaly, is not exactly causal and is hardly inevitable! Nonetheless, the historical illustrations obscure this essential point and hence commit a logical fallacy.

The post hoc approach and cognitive psychology

How is it that discovery could be so unproblematically formulated as a psychological problem? It must be noted again that when the problem of discoveries is posed, it is usually in a post hoc manner. That is, the status of an event *as* a discovery is already settled before the question of how it occurs is announced. Consequently the problem of how discoveries occur or originate or happen becomes equivalent to the question, 'What were the steps which led up to it?' Understood this way, the issue becomes one of making discovery the predicate or outcome of a determinative force or action. The forces examined have been exclusively psychological. For example, before Holton probes Einstein's *Nachlass* for evidence of the latter's confrontation of anomalies and his thematic disposition, it is already taken for granted that Einstein's work was revolutionary. Similarly, before Koestler examines Darwin's autobiography for evidence of bisociation, he already recognizes the value of speciation for biology. Before Kuhn searches the work of Priestley and Lavoisier for evidence of anomalies, he already admits their contribution to the chemical revolution. These cases are candidate examples because *in retrospect* they are examples of discovery. As such the explanations we marshal to account for their occurrence identify immediately the 'creative' forces of cognition.

By contrast, we would find it extremely curious or bizarre to find historians and sociologists of science applying the cognitive explanations of discovery to scientific *failures*. For example, an explanation of the origins of Lamarckian theory, phlogiston chemistry and astrology which resorted to accounts of gestalt shifts, anomalies, ostension, genius, etc., would strike us as inadequate in that such accounts would not uncover what distinguished these accounts from what retrospectively are the true accounts. Indeed, when retrospectively erroneous or bizarre scientific work is examined, the inspection of the psychological forces which produced it tends to focus on the *pathological* aspects of psychology. This is typical of Sigmund Freud's work on Leonardo da Vinci, and Lewis Feuer's work on Einstein's generation.[10] However, we only operationalize this kind of psychoanalytic psychology for the disconcerting 'achievements' in science. When the achievement has proven correct, the hardware of normal cognitive psychology is called into play. In other words, the kind of answer we seek depends on how we pose the question initially. If an achievement is defined as a discovery, the sort of steps which we examine are the workings of creative psychology: gestalt, retroduction, induction, bisociation, etc. However, if the achievement is questionable, bizarre or erroneous, not only could cognitive psychology appear irrelevant to the solution, but the entire software of psychoanalytic psychology could be called into play to make the scientific achievement sensible in terms of childhood experiences, life instincts and death instincts, emotional hydraulics, and ego defences.

Consequently, one of the most decisive elements for the explanation of discovery is the fact that the achievement is defined *as a discovery* from the start. Given that it is a discovery, it is already assumed to be true. Hence, whatever psychological elements are accorded predisposing status, *we are assured from retrospect of their efficacy.* In other words, the social definition of the achievement determines whether the answer lies with Dr Jekyll or Mr Hyde, i.e., the crazy man or the sane scientist.

When we see that the solution is applied under these circumstances, it is no mystery why the necessary condition of mentalistic phenomena appears to be sufficient: the sample of examples is always selected from the winners and the winners are identified, not prospectively, but retrospectively.

The strategic value of psychology

There is a final important aspect to our consideration of the post hoc origins of the problem of discovery. In the previous section we saw that the recourse to psychological accounts was *unproblematic* because these were constructed *after the fact.* However, the question remains, why were the accounts assumed to be *psychological?* A tentative answer lies in a re-interpretation of the basis of Reichenbach's distinction. 'I shall introduce the terms *context of discovery* and *context of justification* to mark this distinction . . . between the thinker's way of finding his theorum and his way of presenting it before a public. . .'[11]

There is some relevant background to this distinction which is pertinent here but which is not widely appreciated.[12] During the 1920s

Mannheim's sociology of knowledge came to some prominence among European intellectuals, including members of the Vienna Circle and their associates. Mannheim presented a refinement of the theory of knowledge originally expounded by Marx in which it was suggested that the ideas in a society represent the interests of the ruling class, i.e. the bourgeoisie. Mannheim held on the contrary that all knowledge did indeed reflect the ethos in which it arose, but that within such an ethos it appeared transcendental. In other words, the truth of a theory was held to be limited to the society in which it arose. This was 'relationalism'.[13]

One of the most notable reactions against such a view was extended by Karl Popper, the philosopher who also has done so much to propound Reichenbach's views among English speaking scientists. These two facts are not inconsistent. Mannheim's theory, with its Marxian overtones, intimates that the conditions under which discoveries are made are overdetermined. This opinion in its most naive form might hold that a scientist is disposed to formulate a discovery on the basis of his historical or *class* background. And since the background of scientists is typically bourgeois, the interests served by the discovery would be a ruling class whose power such discoveries would extend and consolidate – at the expense of the emancipation of the proletariat. If such are the conditions under which ideas arise, might one not expect Reichenbach and Popper to base the doctrine of the separation on political grounds?

One of the consequences of the Reichenbach doctrine is that it is held that, *whatever* the origin of an idea, its truth can be determined independently of it. This is one of the major themes in Popper's *Poverty of Historicism,* when he takes great exception to the *relationalism* of Mannheim and the sociology of knowledge.[14] The veracity of a theory is not justified by changing political and cultural grounds. Hence the context of justification is *apolitical.*

However, the position which argues that ideas bear the mark of their class origins is, according to Popper, an ideology. Consequently, Popper is not merely satisfied with denying the relevance of such factors to the justification of beliefs (which is the primary function of the distinction) but, by focusing on psychology as the domain of the origin of ideas – the context of discovery – it is furthermore implied that the political context is not even the primary context for the *origins* of ideas. In other words, even as politics does not determine the truth of scientific ideas, so too it does not have an inordinate hand in suggesting them. Instead, that which is unique to persons – their psychology, as opposed to their social class, or even social psychology – is the source of discoveries. 'Instead of reducing sociological considerations to the apparently firm basis of the psychology of human nature, we may say that the human factor is *the* ultimately uncertain and wayward element in social life and in all social institutions. Indeed this is the element which ultimately *cannot* be completely controlled by institutions.'[15] It is the unpredictable human factor, which characterizes individual psychology and is the source of scientific ideas and political freedom alike, which Popper and Reichenbach have in mind when the contexts of discovery and justification are separated, and when the question of discovery is posed as a psychological one. All this is lost, however, with the transportation of the Vienna school out of *its* original context of discovery, and its reconstruction in North America. In its new context some of the imperativeness of the distinction has been lost, and the doctrine, now living a life of its own, has become increasingly problematic. Yet the inertia of the school has been uncontestable. Discoveries are intuitively assigned a psychological origin. We shall examine the folk basis of this supposition in the last chapter. However, it would be wrong to think that the various authors find themselves unwaveringly committed to such approaches.

The abandonment of mentalistic accounts

Faced with the above sorts of limitations it is hardly surprising to find the psychological theorists both supplementing their principled models of the discovery process with ad hoc arguments, and abandoning these when it proves convenient. Regarding the first point, for example, Max Wertheimer, having described the symmetry of the law of free fall and its attractiveness to the 'holistic' imagination, suggested that Galileo struck upon it as a lucky guess.[16] Blackwell suggested, as we saw, that Lady Luck often was responsible for the occurrence of discoveries.[17] And all these theorists invoked the notion of genius to explain how, in certain important historical discoveries, the achievements of those particular individuals were made. Aside from the fact that such explanations are usually tautological, the problem in such cases is that these explanations rely in each *particular* historical instance on something which has been struck from the general model.

Lastly, we should note that the mentalistic model, as in Kuhn's case, is abandoned when this proves convenient. In other words, given all the problems associated with an explanation of discovery which relies on arguments about how ideas arise in the mind, we find that an account of this process would not be adequate to explain discovery – *not even for Kuhn*. After identifying 'three legitimate claimants' of the discovery of oxygen, all of whom probably experienced anomalies, gestalt shift, etc., Kuhn goes on to suggest that Scheele, probably the first to isolate and identify oxygen, *can be ignored* because he didn't publish in time.[18] In other words, an account based on the mental origin of discovery is far from adequate for what counts as discovery. Presumably Scheele's work, because it was published late, lacked the necessary novelty or unprecedentedness which the common sense perception of discovery assumes. Consequently, an adequate account of how discoveries occur cannot ever be furnished entirely from inductive or psychological materials. As we shall see in later chapters, this is dramatically confirmed by the case of Mendel before 1900, and the career of *Eoanthropos dawsoni* – the Piltdown Man.[19] Clearly, in these instances the societal context, not the psychology of learning, determines the discovery as a social fact. This is an aspect of discovery we have taken for granted.

This is what the psychology, history and philosophy of science overlook; each is psychological in that each formulates discovery as a mental puzzle to which each offers its own mentalistic solutions: gestalt, ostension, the perception of anomaly, aesthetic motivation, bisociation, retroduction, genius, skill, luck etc. We shall examine the social basis of discovery in the next chapters.

THE UTILITY OF CRITIQUE

Our examination of the mentalistic models is probably as harsh as anything that has been presented. However, we must not lose our sense of proportion in this criticism. There is something enormously beneficial in such a close inspection. In the traditional spirit, 'critique' is not the finding of folly or error; it is the disclosure of the limitation or the contextedness of an approach. The value of this is that it thematizes some area or question which lies beyond the boundaries of the tradition under inspection. In other words, critique is useful in that it delineates what a *more* adequate account would look like; in this sense, a critique is a tacit programme. And surely it must be, for otherwise it would be nothing more than indulgence in follies which, being so awful, would not really merit examination in the first place. The lessons we derive from the mentalistic approaches include the following.

First, a theory of discovery ought to be specifically adequate to explain the occurrence of discoveries and hence must be meaningfully adequate for the social category, 'discovery'. Secondly, the theory must be able to deal with the changes in the social status of these events, and not rely expressly on a post hoc orientation. Consequently, the theory should be geared to the constitutive nature of discoveries, i.e.

their social background or basis, and not assume an inherent naturalistic kernel to which they reduce.

The first attempt to specify a radically *social* model of discovery originated in anthropology. This is our next consideration.

4

The emergence of a social model of discovery

Early in this century anthropologists and sociologists explored a model of discovery which was based on the premise that scientific discoveries cannot be explained by reference to mentalistic phenomena. That this type of explanation is the exact opposite of those we have examined is no coincidence; the original proponents of what we shall call the cultural view of discovery were reacting expressly in opposition to a prominent mentalistic theory based on evolutionary biology. This was Francis Galton's 1869 *Hereditary Genius*. The reaction to this position was part of the widespread rejection of Social Darwinism led by such American social critics as Lester F. Ward, James Mark Baldwin and C. H. Cooley.[1] If social inequality was to be denied a biological foundation based in Spencer's version of evolution, so too was the theory of scientific discovery based on hereditary genius. The keystone of the anthropology and sociology arguments was the historical patterns of multiple discovery. In the early stage of 'the debate' the important figures were on the one hand Francis Galton, and on the other A. L. Kroeber, Ogburn and Thomas and Leslie White.[2] In what follows, we shall trace the development of these ideas from their early proponents to their final expression in the later writings of Robert K. Merton.

MULTIPLE DISCOVERY, CULTURE AND GENIUS

No single observation has proved to be more of an impetus to the sociology and anthropology of science than the observation that the history of scientific discovery is the history of multiple, independent and simultaneous discoveries. That is, the history of science suggests that particular scientific laws and scientific facts have repeatedly been uncovered by different scientists working independently at about the same point in historical time. This observation is the focal point of the famous article by William F. Ogburn and Dorothy S. Thomas, 'Are inventions inevitable?', which contains a list of 148 multiple, simultaneous discoveries in the fields of 'astronomy, mathematics, chemistry,

46

physics, electricity, physiology, biology, psychology and practical mechanical inventions'.[3]

These simultaneous discoveries are cited as evidence that inventions and discoveries occur because of the level of cultural development attained in a society. If a particular discoverer fails to uncover a new law, the society will be none the worse for it, because the history of science indicates that the law will come to light through the work of someone else at about the same time. Hence, 'There is a good deal of evidence to indicate that the accumulation or growth of culture reaches a stage where certain inventions if not inevitable are certainly to a high degree probable, given a certain level of mental ability.'[4] In citing the same passage, Leslie White notably omits the qualifying reference to 'mental ability'.[5] For White, the question of genius is altogether redundant from a cultural perspective. Discoveries occur independently of the I.Q. of specific individuals, and inevitably at a point when the culture reaches a 'critical mass'.

The situation is something like the chain reaction in uranium 235. If the mass of metal is below a certain size a chain reaction is impossible. But when a certain size – 'a critical size' – is reached, the chain reaction is inevitable. Prior to 1843–47, the elements requisite to the formulation of the Law of Conservation of Energy were not available. But, when they became available the interactive culture process made their synthesis so 'inevitable' that it was achieved not one but five times.[6]

White expressly illustrates the *un*importance of genius in his discussion of Dr Urey's Nobel Prize winning discovery of the heavy isotope of hydrogen. The suggestion is not that scientists are of mediocre intelligence, but that high intelligence *per se* is not an extraordinary phenomenon. 'Intelligence of a high order was not essential to the isolation of heavy hydrogen, and we now wish to make this implied conclusion explicit and unequivocal: it could have been achieved by a very ordinary intelligence.'[7] White continues with the suggestion that 'many a household problem – such as . . . opening a recalcitrant jar of pickles – requires as much ingenuity . . . as that required in the isolation of heavy hydrogen'.[8] Therefore genius is an inoperative contingency to scientific success.

The question of genius: biology versus culture

Among the early functionalists, arguments from genius were soundly rejected. Emphasis was placed instead on what Kroeber called the 'superorganism' of human culture. Kroeber, who was the main source for Ogburn and Thomas, and for Leslie White, was apparently responding to the emphasis put on intelligence by the English luminary, Sir Francis Galton.

Galton was certainly one of the most interesting figures of the nineteenth century. His accomplishments were as unusual as they were manifold. Having explored sections of the uncharted continent of Africa, he returned to England where he pioneered the field of 'biometrics', the application of statistics to biology, on which he based his theory of hereditary intelligence. In his mania for measurement, Galton did research into such things as the average number of brush strokes in portrait paintings, and the tedium of sermons educed by the fidgets of the congregation. In the same vein, he published the results of a statistical study showing the inefficacy of prayer in currying divine favour. One of his more enduring discoveries was that human finger prints, far from identifying individual *races* of men, were unique to specific *individuals:* the implementation of this fact by Scotland Yard proved to be of inestimable aid in the control of crime.[9]

For our purposes, Galton's most noteworthy research concerned the distribution of human talent in society. In *Hereditary Genius,* Galton observed that the most notable figures in all fields – jurisprudence, politics, music, science, literature, athletics, etc. – typically belonged to families which showed other evidence of similar talents. For example, the Bernoulli family produced eight remarkable mathematicians over a period of a century and a half. Likewise the Bach family produced several important musicians over two centuries. Galton himself was cousin to Charles Darwin and grandson of Erasmus Darwin and descended through his mother from the talented Wedgwood china potters. Through the method of correlative coefficients, Galton determined that these patterns of genius did not occur by chance.

The correlative coefficient which was Galton's most significant single contribution to science is a statistical description of the degree of association of two variables. This method, pioneered by Galton and perfected by his friend and student Karl Pearson, became the methodological foundation for biometrics. Galton conducted extensive measures of the heritability of physical and mental traits, and the tendency in the offspring to revert or 'regress' to the population average for such traits. In *Hereditary Genius,* he argued that the genius of the families or clans he studied was organically based, and was consequently perpetuated over generations by biological inheritance.[10] It followed that the great periods in science resulted from the achievements of researchers whose genetic constitutions were especially adept at solving scientific problems. Similarly in music, jurisprudence, etc., the great contributions were the works of hereditary genius.

By implication, great periods of development in civilization were tied to the successful breeding of extraordinary individuals; and by the same token the decline of great civilizations resulted from their inter-breeding with duller races. Regarding the golden age of Athens, Galton suggested:

In a small sea-bordered country, where emigration and immigration are constantly going on, and where the manners are as dissolute as were those of Greece in the period of which I speak, the purity of a race would necessarily fail. It can be, therefore, no surprise to us, though it has been a severe misfortune to humanity, that the high Athenian breed decayed and disappeared.[11]

There was naturally great exception taken to such views, as there was exception taken to the related views of the Social Darwinists. Indeed, Cooley, who was noted for his opposition to Social Darwinism, also published a persuasive rebuttal of Galton.[12] However, the most enduring challenge to Galton was given by A. L. Kroeber, who ardently rejected the determination of the social world by the organic. According to Kroeber great periods of history emerged as a function of cultural development. Nor was Kroeber prepared to settle for the familiar nature–nurture synthesis. He reasoned that 'Organic evolution is essentially and inevitably connected with the hereditary process [but] the social evolution which characterizes the progress of civilization, on the other hand, is not, or not necessarily tied up with hereditary agencies.'[13]

Arguments of hereditary agency, especially in the form espoused by Galton, were not unproblematic in their reduction of cultural forms to psychologically expressed organic conditions. Galton's statistical method of study revealed that talent occurred with levels of significance far greater than average in particular families; but such a result could be expected to occur randomly in certain numbers of families in the general population. In other words, Galton's sample tended to be unrepresentative.[14] The argument was further hampered by the fact that even if his samples had been representative *no* allowance could be made for the effect of environment in the perpetuation of talent. As Cooley's article pointed out, every genius Galton discussed, whether born into rags or riches, had benefited perceptibly from formal or informal education. Consequently Galton's position was highly problematic.

What is surprising in Kroeber's attack on Galton is that he failed to explore these inherent weaknesses in the organic position. Instead, the key element in Kroeber's argument was his exposition of the historical patterns of discovery in science – patterns which forcefully un-

dermined the value of individual intelligence. As we shall see, this
tactic was as overzealous on the cultural side as was Galton's on the
organic side. These discoveries, according to Kroeber, follow patterns
of cultural 'maturation': the evidence, of course, a full page of multiple,
simultaneous and independent discoveries.[15] To argue that these patterns of simultaneity were coincidental expressions of genius would
be presumptuous. On the contrary, discoveries occur because their
time has come. For example, regarding the mutual independent discovery of speciation by Darwin and Wallace, Kroeber suggests that it
is inconceivable that the same idea should have come to each as a
matter of 'pure chance'. Kroeber continues, 'The immediate acceptance of the idea by the world proves nothing as to the intrinsic truth
of the concept; but it does establish the readiness of the world, that is,
of the civilization of the time, for the doctrine . . . the enunciation
seems to have been destined to come precisely when it did come.'[16]
Concomitantly, the failure of Mendel to gain social recognition for his
contribution to heredity is 'an instance of the inexorable fate in store
for the discoverer who anticipates his time'.[17] And what of the eventual
rediscovery of Mendel's paper by DeVries and Correns? 'It was discovered in 1900 because it could have been discovered only then, and
because it infallibly must have been discovered then.'[18] In other
words, the discovery occurs 'genetically' with the ripeness of time.
That its time has come is evident from its appearance; the discoverers
are simply midwives to the 'march of history, or as it is current custom
to name it, the progress of civilization'.[19] Needless to say, this reasoning is logically circular.

It has been pointed out earlier that the use of arguments of genius,
insight, and luck to explain the occurrence of discoveries when the
only evidence of the genius is the outcome which it is used to explain
is tautological. Specifically it was noted that this type of argument is
used, as in the case of Hanson and Blackwell, to make sense of the
relative success/failure of competent rivals. However, in the present
writings the use of arguments from genius is seen as *redundant* or
irrelevant. Notes White, 'We do not need to take the brains of men
into account in an explanation of mathematical growth and invention
any more than we have to take the telephone wires into consideration
when we wish to explain the conversation it carries . . .' And what is
the proof of this? White continues: 'Proof of this lies in the fact of
numerous inventions (or 'discoveries') . . . made simultaneously by
two or more persons working independently.'[20] Consequently genius
has no bearing on the pattern of discovery in science. However, the
citation of the list of multiples has replaced the circular argument from

genius with the equally circular argument from culture; to say that a change occurs because its time has come ascribes to temporality, which is a property of events, a causal status which more properly belongs to those events which are implicated by such references.[21] Thus by imputing to culture the model of the 'superorganism', we seem no better able to fathom the cause of discovery in this outgrown genetic form than were the culturologists when they were confronted with Galton's organic position.

To make matters worse, White admitted that one could not always specifically determine what conditions governed the origin of a cultural critical mass. Though in the main 'the culturological interpretation . . . tells us, how, why and when a genius will appear',[22] still, some events belong to the 'fortuitous course of history',[23] that 'chronological series of events each of which is unique'.[24] For example, no one could predict that Booth would kill Lincoln; that event was historically unique. However, the *only* evidence of those other, predictable and inevitable events of cultural evolution is the list of multiple discoveries and inventions. Consequently the culturologists have an air-tight case: when genius or individuality or serendipity does occur, this is historical uniqueness; otherwise, social and scientific change is culturally determined. In other words, anything that cannot be predicted or explained with reference to the circular argument from culture is confidently chalked up to historical singularity.

The strength of the argument from culture lies in the citations of multiple discoveries. However, the reliance by these authors on such lists is rhetorical. Kroeber's initial identification included some sixteen multiples; Ogburn and Thomas expanded this list probably on the suggestion given by Kroeber that 'a volume could be written, with but few years' toil, filled with endlessly repeating, but ever new accumulation of such instances'.[25] No such volume was ever written; the tradition settled instead for the appendix to the brief Ogburn and Thomas article which identified 148 simultaneous discoveries. However, the significance of the list was never explicated in detail. Specifically, it has never been shown what proportion of the discoveries in science have been multiple and what proportion single. Indeed it is unclear if it is even possible to count discoveries like so many chickens in a coop. Nor is it clear at what point the relative proportions are significant. The distinction between historical singularities and cultural determinism compounds the matter even more, for the culturologists could dismiss all counterfactual instances as 'unique genius'. The list inspired by Kroeber has taken on an unchallenged superorganic significance in the sociology of science up to the present time. Indeed, that list of

148 has swollen out of all proportions in Merton's imagination when he suggests, 'The pages of the history of science record thousands of similar discoveries having been made by scientists working independently of one another.'[26]

MERTON ON MULTIPLES, GENIUS AND PRIORITY STRUGGLES

The writings which concern us here appeared in 1957 and thereafter, and constitute what Norman Storer calls 'the heart of the Mertonian paradigm'.[27] We shall deal with their content in three parts: Merton's hypothesis that all discoveries are in principle multiple discoveries, Merton's functional interpretation of genius, and Merton's famous discussion of the normative origin of priority disputes.

The case for discovery as a multiple event

The central paper here is entitled 'Singletons and multiples in science'. Like Merton's other writings of this period, it is grounded thematically in the Ogburn and Thomas list. The central observation is contained in the following: 'the hypothesis states that all scientific discoveries are in principle multiples, including those that on the surface appear to be singleton'.[28] To support this position Merton offers 'ten kinds of related evidence'. This evidence is *not* a survey of the actual history of science providing descriptions of the frequencies of multiples and singletons: in light of Ogburn and Thomas, such a task apparently seems redundant. Rather, Merton examines numerous cases in which multiples have been experienced as the norm by the researchers themselves. Some of these points have been clustered because of their redundancy.

i. first is the class of discoveries long regarded as singletons which turn out to be rediscoveries of previously unpublished work.[29]

For example, numerous discoveries recorded by Cavendish in his notebooks on chemistry and electricity were published posthumously; in the meantime that work was duplicated by some dozen other researchers. The same thing occurred with the tardy publications of Gauss in mathematics.

ii, iii, iv. Second . . . in every one of the sciences . . . there are reports . . . stating that a scientist has discontinued an inquiry . . . because a new publication has anticipated his hypothesis.[30]

Consequently though a single publication may come to light, this does not mean that only one party was working on the problem. The third class identifies those scientists described in the above predicament, but who finish their work, publish it, and cite the prior success of the newly discovered rival. This category is available because the scientists have published their work, even though it has been anticipated. However, 'many scientists cannot bring themselves to report in print that they have been forestalled'.[31] Ergo, there is a possible fourth category which has been forestalled but about which no announcement has been made. Ironically, the absence of any instances is taken to support the claim.

Other evidence that discoveries are in principle multiples is derived from the observation that:

v, vi. seeming singletons repeatedly turn out to be multiples as friends, enemies, co-workers, teachers, students[32]

inform the researcher of prior successes or rivalry on the same subject. This point is really just an instance of the first point (i.e. that apparently unique work is only a duplicate), as is the sixth point, which suggests that the news of rivalry or duplication may be communicated to the researcher by his lecturer.

vii. Scientists who are diverted from their plans 'by the authorities' may later be succeeded by others working independently on the same programme. Hence for every singleton, 'there but for external interference', goes a multiple.

viii, ix, x. The actual behaviour of scientists repeatedly illustrates their conscious preoccupation to publish first, and to defend their priority. The defence of another's priority on his behalf is Merton's ninth point; and the willingness of certain institutions to accept sealed announcements to assure a researcher's priority is Merton's tenth point. These last three items presumably illustrate that researchers are preoccupied with the problem of rivalry or priority, because of their personal knowledge of it.[33]

These several points provide evidence that multiples have been identified frequently by scientists, and that scientists consequently show some concern for priority. However, these points cannot be said to bear out the hypothesis that discoveries are *in principle* multiples. They establish that knowledge of multiples is common, if by common we mean present in the literature which Merton cites favourably. However, there is no guarantee of the representativeness nor exhaustiveness of his citations. Merton's exposition is more speculative than logical or thorough. We shall examine the list more systematically at a later point.

Merton on genius

Although Merton speaks about constructing a 'theory of scientific discovery' he in fact develops two theories. His more noted theory concerns the normative constraints governing the behaviour of scientists competing in the same research. The other theory is itself quite vague in Merton's work, although it is more exactly what we would recognize as a theory of discovery *per se* – that is, an explanation of the conditions under which discoveries emerge. The chief elements in his account are conditions of *cultural maturity plus* conditions of *genius*. Far from being antithetical, Merton implies that *both* contribute to the occurrence of scientific discoveries.

Genius is defined as follows: 'The individual of scientific genius is the functional equivalent of a considerable array of other scientists of varying degrees of talent. On this hypothesis, the undeniably large stature of great scientists remains acknowledged.'[34] Merton is responding here to the observation that some historically renowned scientists like Galileo, Newton, Gauss, Faraday, Maxwell, Hooke, Cavendish, Lavoisier and Priestley were each involved in several cases where their works were independently rediscovered by numerous others. Hence the one genius was functionally equivalent to his several successors. However, if all discoveries are in principle multiple, and if genius is reckoned as the functional equivalent of an array of others, *any* discovery is proof of genius, for a discovery is by definition a multiple; and hence any discoverer is equivalent to numbers of others. Similarly, the more discoveries, the larger the number of functional equivalents, the greater the genius. Therefore, the efficacy of genius in scientific discovery is guaranteed by definition. This hardly constitutes a rehabilitation of genius.

The error is exposed by the following. If for any discovery there are N researchers working simultaneously, one of whom eventually succeeds before the others, do we conclude that the successful one was functionally equivalent to $N-1$ other talents? Merton seems to suggest yes, but this is false. We can only say that he or she is functionally equivalent to *any one other* researcher. Some researcher is necessary; any one is sufficient. Consequently the genius is not greater in stature, but only prior in time to his rivals.

If a researcher made several discoveries, is his mental stature cumulatively enlarged? As far as the theorist is concerned, each further discovery is evidence of greater functional equivalence, that is, greater evidence of talent. This again puts the theorist in the circular position of deducing the preceding talent from the consequential discovery where the sole evidence of the former is the latter. This logic is fallacious.

By the exclusion of genius we are left with the sole argument from culture. This, however, is not discussed at any length by Merton. He merely reiterates the Ogburn and Thomas position that 'innovations became virtually inevitable as certain kinds of knowledge accumulated in the cultural heritage and as social developments directed the attention of investigators to particular problems'.[35] Consequently the force of Merton's sociological account of the process of discovery turns on the adequacy of the explanation offered by Ogburn and Thomas, Kroeber and Leslie White regarding the inevitability of cultural change and especially the changes in science. However, Merton's unique contribution to the sociology of discovery is his explanation of the phenomena associated with researchers' rivalry over competitive contributions, and the explanation of such events in terms of the normative order of scientific institutions. This research has proved to be his most fruitful in that it initiated a wave of investigations into numerous aspects of the reward structure in science.

Priority struggles in science

Merton's study of priority struggles suggests the following: the history of science is characterized not only by ubiquitous incidents of multiple discovery, but by repeated quarrels among scientists regarding their claims to priority in certain discoveries and, consequently, their relative entitlement to social recognition.[36] Since Merton holds all discoveries to be in principle multiple discoveries, this issue of priority is enormous. In the history of science, there is no dearth of famous disputes. Galileo battled Baldassar, Father Horatio Grassi, Simon Marius, and a host of others who, Galileo claimed, had stolen his ideas. Newton quarrelled with Hooke over optics, and with Leibnitz over the invention of calculus. Cavendish was involved in disputes with Watt and Lavoisier over the compound nature of water. Adams and Leverrier squabbled over the claim to the discovery of Neptune. This list goes on and on.

Merton dismisses the explanation that the priority struggles are motivated by egotistical claims to the discovery, or the exceptionally aggressive personalities of some scientists. He notes that quarrels are sometimes carried on, not directly by the principals themselves, but by the friends and benefactors of those parties, as in the case of Cavendish and Watt. According to Merton, this pattern suggests that com-

munity norms of originality and humility or disinterestedness are central here: 'It is these norms which exert pressure on scientists to assert
their claims, and this goes far toward explaining the seeming paradox
that even those meek and unaggressive men, ordinarily slow to press
their own claims in other spheres of life, will often do so in their scientific work'.[37] Such activities are motivated by the norm of originality.
Originality, which is at the heart of a discovery, is rewarded institutionally by recognition. 'Recognition for originality becomes socially vali-
dated testimony that one has successfully lived up to the most exact-
ing requirements of one's role as a scientist.'[38] Originality is rewarded
above all in the form of eponymy – or naming the law or phenomenon
after the discoverer. 'Eponymy is . . . that most enduring and perhaps
most prestigious kind of recognition institutionalized in science.'[39]
However, other rewards include the receipt of major awards, most especially the Nobel Prize.[40] Also, in some countries leading scientists
have been ennobled: for example, Newton was knighted, and Sir William Thomson was re-titled Lord Kelvin. There are many lesser forms
of recognition such as community respect, institutional status, membership in prestigious bodies, honorary degrees etc. The existence of
such reward structures regulates the adherence to the norms of originality.

However, originality is not the only norm; scientists are also guided
by a concern for 'disinterestedness, universalism, organized scepticism, communism of intellectual property, and humility'.[41] At times
these norms may be inconsistent. Indeed there is a tension between
originality and humility which leads scientists to 'insist on how little
they have been able to accomplish' and 'creates an inner conflict
among men of science who have internalized both of them and generates a distinct ambivalence toward the claiming of priorities'.[42] This
value conflict is used to explain the mixed feelings experienced by
scientists embroiled in issues of priority, and to explain why on occasion certain individuals would allow their grievances to be expressed
only through others.

However, the only example given of the mixed feelings experienced
by scientists is the case of Darwin and Wallace. It cannot be doubted
that Darwin (*not* Wallace) felt that 'all his originality was smashed' by
Wallace's presence; yet it cannot be argued on the other hand that
they were involved in any struggle or quarrel over their separate originality, for no such claims were ever made. By ignoring such details
Merton overextends his case for the efficacy of the normative order.

He notes that institutionally greater rewards are accorded *more* to
originality than humility. This explains the occurrence of priority dis-

putes in spite of the negative feelings which these might entail. It also explains the occurrence of various forms of scientific *deviance* which are thematically tied to this bias in the reward structure of science. Following Merton's familiar typology of deviance, fraud and plagiarism are examined as expressions of deviant means oriented towards institutional ends. Other responses include the stock in trade of normal science: symbolic originality ('publishing'), retreatism and cynicism. These alternative responses to originality conclude Merton's investigation of priority.

This investigation was thematically extended in several studies conducted by students of Merton, most notably, H. Zuckerman, Jonathan Cole and Stephen Cole, into the general features of the reward system in modern scientific communities.[43] The exact dimensions of that work does not concern us here, for these studies have moved further afield from the structure of discovery than the initial investigations on which they are based. They stand on their own merits.

As noted earlier, the study of priority does not expressly constitute an explanation of discovery. It is founded on 'the strategic fact' of multiples,[44] and the concomitant observation that these are sources of disputes for scientists. It proposes to explain the disputes often associated with multiples with reference to the norms of originality and humility. However, a number of counterfactual instances raise doubts about the real origin of priority disputes. As Merton notes, Darwin and Wallace's co-emergence did not lead to any quarrel over priority. 'Darwin and Wallace tried to outdo one another, giving credit to the other for what each had separately worked out.'[45] However, this is taken to be the exception, rather than the rule, a conclusion very difficult to sustain without a representative examination of priority claims. Of course, there is none. As for the ubiquity of humility, Merton cites several instances of scientists who expressly flaunt this institutional value – Watson and Crick, Norbert Weiner, Hans Selye, Herbert Spencer, Linus Pauling and P. B. Medawar, to name several. Again these are the exceptions rather than the rule, and again this is plainly unsubstantiated.

The fact is that the bulk of the disputes revolved around questions of *fraud* and *plagiarism*. Galileo attacked his rivals because he suspected they had somehow stolen his ideas. Merton mentions in passing sixteen or seventeen priority disputes, without examining any in detail. In light of the ambiguity of what constitutes a priority struggle, and in the absence of a systematic description of their frequency in the historical calendar, the entire matter is far from settled. Do the disputes derive from the normative originality? Is originality a specifically 'institutional' norm? Or is it an epistemic condition for discovery? Is it not clear that some disputes originate over the suspicion of having been robbed? Or over charges of having robbed someone else? Which of these is operative in the multi-faceted history of priority disputes? These are questions whose answers turn on the details of the actual situations.

As it stands, the explanation based on the institutional norms of science is not overwhelming or compelling. When disputes associated with rival discoveries fail to occur, an absence of conflict is unexplained. Presumably it has something to do with what Merton calls the personal 'noblesse oblige' of the individuals, as in Darwin's case. This, however, seems to be an ironic substitution for the earlier vacillations found in Leslie White between historical singularities and cultural determinations. When events occur as predicted, this is normative behaviour; when not, this is noblesse oblige.

As we can see, the identification of the social dimensions attendant on the institution of discovery opens up more issues than the normative model can account for. The issue of priority is not entirely separate from the historical pattern of multiple discoveries, for we see that the challenge by one scientist that another has stolen his ideas is in fact a challenge to the social basis or social definition of the other's discovery. If others have stolen my ideas, this is not evidence of multiple discoveries, but fraudulent claims to discovery – which are not discoveries at all. In other words, priority disputes may often be cases of competing definitions of the situation. In sum, Merton's identification of the institutional aspects of science turns our attention to the behaviour of scientists which in turn indicates that it is *their* work to mediate the status of an event, i.e. to determine whether some achievement is a discovery, a replication, a flatulent doctrine, a pious belief, an intellectual theft or whatever. While it may very often be the case that certain researchers do independently discover the same law, the recognition of this fact entails an assumption about the likelihood of such a coincidence, the identity of the contributions, the motivation of the researchers, etc. Such being the case, *a sociological account of discovery turns on the processes whereby certain events are determined socially to be discoveries* – that is, are defined as discoveries. This suggests that certain of Merton's basic concepts can fruitfully be re-interpreted to establish the viability of the idea that discoveries are socially constructed.

INSTITUTIONAL NORMS OR COMMON SENSE KNOWLEDGE OF DISCOVERY?

The following interpretation is based on the assumption that the actual behaviours of scientists are clues to their understandings of discovery, and furthermore that knowledge of the latter is a prerequisite for an adequate sociological account. The emphasis here is not on the institutional norms said to govern the behaviour of scientists, but on the scientist's own common sense knowledge of discoveries. Three of Merton's observations will be re-examined: that discoveries are in principle multiples, that their achievements are motivated by normative originality and that social recognition rewards originality.

Discoveries are in principle singletons

Even a cursory re-evaluation reveals that Merton's data, which were offered to establish the omni-presence of multiples, can be interpreted by attending to the understandings of the scientists themselves, to show that discoveries in science are unique and singular, that is, *that discoveries are in principle singletons*. The anxiety which is exhibited by the researchers in their rush to have new findings published, in their attempts to have sealed disclosures of their discoveries deposited with scientific academies, to withhold news of imminent progress from their rivals, and to terminate research when they learn of being forestalled, all indicate that the scientists themselves understand discovery to be *in principle* a singular event. Their behaviour is sensible only if this is true. If on the contrary it were assumed that a number of workers could arrive at the same law without this affecting the social dimensions of all the announcements, we would not witness this preoccupation with the singular character of discovery. However, it is common knowledge that an announcement which merely re-iterates what has already been noted is 'late'; or in science it is a 'replication'. In other words, *the discovery changes the order of events into which it is announced rendering the subsequent context different for later identical announcements.* This points to the common sense definition of uniqueness, singularity or novelty in scientific discovery.[46]

Singularity or normative originality?

The perception by scientists that discoveries are singular events is tied, not to the institutional norm of originality (whose efficacy *per se* is very questionable) but to the common sense judgements about the *novelty* of an achievement. For example, the mathematician who devises an original solution to a long standing problem, yet who is forestalled in his publishing, is nonetheless original in his efforts. His solution is original inasmuch as it originated solely from his own thinking. However, his publication in the community will lack novelty; in other words, his work might constitute a personal triumph, but the announcement as far as the community is concerned will not be news. In the context of the community, novelty as opposed to originality is at the root of discovery.

Stated otherwise, from the point of view of those who recognize that an achievement is a discovery, it is in the nature of the phenomenon that it be unprecedented. It is therefore a *grammatical* requirement of the meaning of the event that its annunciation be unique, original or unprecedented. In this sense, the condition of unprecedentedness or singularity in *the definition* of an event as a discovery is more fundamental than the institutional norm 'to be original'. If the researcher has made a discovery, this entails originality by definition. However, the unprecedentedness of the achievement can only emerge against a background of a socially sanctioned recognition.

Social recognition as constitution and as reward

The concept of social recognition is central to Merton's position. However, for Merton, recognition is equivalent to *social reward*. It is this interpretation of recognition which has directed his paradigm to studies of social stratification in science. A discoverer is 'given recognition' via a place in the scientific hall of fame, membership in a royal society, academic status, community respect and philanthropical money. All these are ways of expressing 'recognition'. In this conception, recognition centres on the person involved and his share in the fruits of success. Consequently, Merton has some difficulty separating the conception of recognition as individual fame motivated by egoism, from recognition constituted purely by acknowledgement of the theory. 'Sometimes . . . the need for recognition is stepped up until it gets out of hand: the desire for recognition becomes a driving lust for acclaim (even when unwarranted), megalomania replaces the comfort of reassurance.'[47]

Presumably the norm is usually a psychologically more modest event located somewhere between pure vanity on the one hand and social indifference on the other. However, no matter how motivated the researcher is for a community approval, the recognition at issue here is still *psychologically* grounded. Although his concern for recognition is occasioned by the reward structure in science, such recognition is figured in a desire/reinforcement matrix which assumes a social approval model of behaviour in which choices are explained in terms of the gratification which these guarantee for the individual. This type of focus is an ingenious explanation of the *unconventional* attempts to draw social approval witnessed in the accounts of fraud and plagiarism on the one hand, and the psychologically antithetical situations of retreatism and cynicism on the other. However, these are extreme cases of a *psychological* adaptation to the institutional opportunities of science; and Merton's approach is more than adequate to understand them.

However, I would suggest that there is a more strictly sociological sense of recognition in which the achievement is itself the source of the reward. That is, the motivation to uncover a natural law is completed by the resolution of the problem or investigation in a successful outcome. Recognition in this sense is equivalent to the *acknowledgement* that a proposal is true, or that an objective was met. *This recognition consists of our apprehension of the kind of phenomenon which the achievement constituted.* The issue from the sociological perspective is *not* the degree to which the individual is rewarded for doing a job well – but whether the individual's efforts are seen to constitute a 'job' or 'achievement' at all! In other words, the recognition is recognition *that* a discovery has indeed been made, *versus* the reward given to its originator.

In the case of Columbus, which we shall examine in more detail in a later chapter, we are not especially concerned with the personal honours bestowed on 'El Admirale' on his return to Spain; the more fundamental recognition was that new land masses existed and that this came to be apprehended as a virtual discovery throughout Europe. Columbus' social reward only followed this more basic recognition. That is, because this accomplishment was recognized, the concrete rewards could follow.[48] While we tend as members of society to intertwine these two things in our zealous celebration of the events, analytically two separate matters are involved: one concerns the reward structure of science, the other concerns its cognitive structure.

This same issue is nicely borne out by the observations of Dorothy Thomas in her construction of the list of multiple discoveries.[49] Whereas one would have thought that her task of listing the multiples would have been relatively uncomplicated, this did not prove to be the case. The historical record was plagued with several major difficulties. It included cases of ostensive multiples in which one of the parties accused his contemporaries of having stolen his ideas, and cases in which, in retrospect, certain of the claims turned out to be illegitimate. It included cases in which the dating was ambiguous and cases in which the contributions, though concurrent, were radically different in quality and completion. It included as well cases of simultaneous discovery whose significance was not locally recognized, but was only reconstructed from the benefit of hindsight. These conditions are of interest here because they substantiate the present conception of discovery: that discoveries do not occur in a naturalistic manner, but are socially constructed. Thomas' predicament recreates on a miniature scale the common sense issues which attend the ordinary member of society and his or her recognition of discoveries. Her dilemma stems from the question of which achievements, in view of all the circumstances, can *meaningfully* be recognized or acknowledged as real discoveries.

5
Discovery as meaningful action

THE SOCIAL BASIS OF DISCOVERY

Generally speaking, most research in the sociology of science can be classified into one of two main categories. First, there is work which evaluates the impact of science on society, thus treating discoveries or other aspects of scientific culture as independent variables. Such studies include the effect of technologies on social and economic modernization, the conflict of science and traditional beliefs, the impetus of biometrics on the eugenics movement in British politics, recombinant DNA research and civil safety, technologies and environmental pollution, etc. However, the more common work found in the sociology of science evaluates the effect of society on science. The strategy in this work has been to assign certain scientific developments to 'external' social causes. Merton's research on the Puritan ethic and the rise of scientific institutions in England is an exemplary case in point. However, this approach would include other things like the effect of the military-industrial complex on the control of research, political sponsorship of scientists and their institutions, social position and the adoption of the scientific mentality, and the stratification system and its effect on productivity.

This latter approach, while being eminently worthwhile, does have a certain liability. It skews the topics of the sociology of science and knowledge to the marginal, unusual or 'extra-scientific' dimensions of science. As David Bloor points out in *Knowledge and Social Imagery,* the sociology of science runs the risk of becoming a sociology of error – for we tend to reserve sociological causes for the out-of-the-ordinary dimensions of science. Notes Bloor, one tends to think 'that nothing makes people do things that are correct but something does make, or cause, them to go wrong'.[1] Consequently, sociology is called upon to explain why Mendel was 'overlooked', why Galileo and Giordano Bruno were persecuted by the Church, why Lysenko prevailed in Russia at the expense of Mendelian genetics. In these cases, the social basis of scientific phenomena is considered only in a conspiratorial or political sense, that is, as 'social influence'. However, the programme developed over the last decade by a number of British sociologists of science calls for a substantively scientific account of science, that is, a programme which examines the actual theories, decisions and practices of working scientists as topics worthy of study in their own right, no matter how politically uneventful, and understandable in their own terms, no matter how prosaic.[2] This focus on the 'cognitive' elements in science represents an expansion of the sociology of science beyond the 'normative' analysis of scientific communities initiated by Merton.

The present work is consistent with the cognitive school. It takes the position that we should articulate the social basis of discoveries not through an examination of 'social influence', but through a phenomenological approach that posits that all phenomena endemic to scientific research are socially constituted and identified – not in the sense of being 'swayed' by extrascientific facts, but in the sense of being constructed by members of society as 'scientific' in the first place. This is true above all with discovery. In this chapter I will outline a model of how members of society confer the status of scientific discovery on candidate events by virtue of criteria of intelligibility learned as part of our common stock of knowledge about the world. I will then consider certain problems which possibly plague the model, namely, the problem of relativism associated with the phenomenological perspective, the problem of the causality of the model and lastly the way in which this conceptual model relates to the details of actual settings.

However, before we discuss scientific discovery, it will be useful to examine briefly another 'constitutive' or 'phenomenological' investigation which displays a structure somewhat parallel to discovery and which largely animated this work.

THE COGNITIVE BASIS OF LEARNING

The research which is pertinent here was conducted by Robert Mackay and Aaron Cicourel on the processes involved in the assessment of reading skills.[3] Traditional pedagogy has assumed that learning is based on intelligence, and that intelligence reflects the ability to reason, which is itself based on some mixture of genetics and experience. For our purposes, the hallmark of this understanding is the ascription of learning to some innate abilities of the individual. Common sensically, learning is measured by the degree of success that one experiences in performing certain novel tasks and recalling certain past events. Mackay's ethnographic work suggests, on the contrary, that learning consists in the institutional ascription of success whereby certain preferred responses are selectively ordered and identified as learning achievements to the exclusion of other meaningful performances.

The point of departure for Mackay's inspection of classrooom activity is the developmental conception of schooling. This is a folk understanding which pervades Western teaching institutions and which holds that the child in society is an underdeveloped, incomplete or immature form of the adult. Schooling then, is assumed to be the process of delivering children from their role of social incompetence, if not to the role of adult, then at least to the role of adolescence (i.e. the next best thing).

Mackay argues by contrast that school children belong to a separate cultural world in which they are highly competent as interpretive members; with other children, their performances and abilities are an assured and accountable matter. This is the theory of the children's culture. According to this understanding, 'maturation' or 'development' is the adult culture's gloss for the passage from the children's culture to adult culture. The children's culture is, then, a qualitatively different sort of culture, as opposed to a quantitatively incomplete one. Consequently, what knowledge a child can be said to accumulate and learn is measured from the perspective of something like a 'foreign culture' whose methods of investigation (for example, institutional tests, teachers' perspectives) overlay a blueprint of adult expectations on a culturally different field of responses and interactions.

More specifically, Mackay argues that the methods of the school teacher for testing the recall of the contents of a story circumscribed the dialogue with the child in such a way that the child's particular interpretation of the events correctly addressed the questions which the child had understood to be asked, though the answers were inconsistent with the teacher's interpretation of those events.[4] The child's score consequently suffered. Likewise, *warranted disagreements* on objective measures of reading skills were coded as 'erroneous responses'. All this led to a certification of incompetence. However, this failure to learn was provided for in a more general way by the maturational or developmental hypothesis under whose terms such a failure could readily be anticipated by the 'incomplete' status of the child *vis à vis* the adult. Given the understanding that two cultures are present, then we see that what is called *learning* is not a psychological or mentalistic phenomenon pure and simple, but is a feature of the institutionally organized attribution of success and failure based on preferred responses.

This conception of learning provides a useful parallel to the social basis of discovery. It suggests that we study discovery as a feature of the accounting practices by which it is rendered possible in the first place. Although as folk members of society we automatically interpret individual discovery or learning as the outcome of a motivated course of inference, sociologically we must consider the cognitive and empirical grounds in terms of which such an achievement is figured. From this position, cleverness in school is understood, not as a function of innate mental powers, but as a function of the context in which the achievements associated with cleverness are made accountable and remarkable – likewise for the scientist. His discovery must be inspected not for its content or psychological origins, but for the context which makes it a possibility or a candidate in the first place. This candidacy status of events is what I mean by the social basis of discovery.

Common sense understandings regarding scientific discovery

That discovery and learning have a social basis is indicated in a general manner by the linguistic or epistemic limitations associated with these notions. Because of our mastery of language and our common stock of knowledge, we know that only certain types of events can or cannot have candidacy status in that they can or cannot be designated in common usage as learning or discovery. For example, it is possible to learn to play the oboe, or to learn the theories of Newton, or to discover new stars, asteroids or U.F.O.s. On the contrary, what sense is there in the idea that one can learn to dream, or to sleep, or to learn to faint or to free associate, or to focus one's eyes? Certainly one can learn to control the *timing* of one's sleep, and certainly one can pretend to faint, and learn to relax and concentrate on ideas; but these are different matters. The sleeping, fainting and dreaming are not voluntary, and hence they are not achievements, and hence they are not the kind of events which could be the outcome of a motivated course of action designed for their attainment. In short, it makes no sense to speak of 'learning' to sleep, etc. These events are not appropriations; they 'occur naturally'.

Similarly I cannot learn steps which tell me how to make a specific discovery like the steps which describe how to make beer. If such steps were known beforehand (Faust notwithstanding), the outcome would not be a discovery, but a replication. So too, in science I cannot knowingly discover what others already have reported as true. Each of these would undermine the essential *novelty* associated with discovery. Also I cannot claim to have discovered something which I know to be false, or which does not exist. Such claims would belie the essential *veracity* associated with discovery. Psychic researchers are finding considerable resistance to their achievements on just such grounds. In the same view, I cannot claim to discover what appears to be an ontological impossibility, for example, an unconscious toothache.[5] As a scientist, I cannot claim to discover the ultimate meaning of existence; such a claim would be outside the realm of science, and would presumably be sensible only in speculative philosophy. I take all these claims to be prima facie self evident to other members of this culture. These understandings constitute part of our society's folkways regarding the social facts of discovery.

Therefore we see that not any event is a candidate for discovery. Before the acknowledgement of any particular scientific achievement, finite background understandings are operating to ensure the candidacy of certain types of events at the expense of others. In other words, we operate with a criterion. Hence, to speak of some event as a discovery assumes that the conditions for discovery are present.

In earlier chapters it was noted that the conditions specified for the occurrence of discovery focused on mentalistic phenomena whose identification could not provide for the distinction between discovery and learning. In this chapter we have seen that learning itself is provided for by certain institutional understandings, most notably the maturation hypothesis which lends to certain events their sense as episodes of learning (and as failure to learn). In other words, one's performance on a test is deemed an accountable measure of what-one-has-learned because it is situated in the midst of definite courses of action whose make-up is commonly recognized as 'doing' learning. This seems self evident. However, imagine how a fictional anthropologist from the Trobriand Islands might report the practices of our society. The people of North America segregate large numbers of their children in institutions with small numbers of adults, who make them read nonsense stories, which they then force the children to recall without written aids. Such mortifying trials are consequently used to determine the child's social standing. This custom is called 'learning' and 'having a test'. Clearly this is not our intuitive interpretation of these events. However, the point of this characterization is to suggest that the various courses of action which are concretely associated with learning and school performance take on their value 'for the locals' by virtue of a background knowledge of what constitutes learning. Primary in this understanding is the maturational hypothesis. This provides an overarching understanding in terms of which the various episodes coalesce into a unifying motivational matrix. Our familiarity with what is called learning allows us to perceive the continuity between the *reading* of stories about fictional characters such as

Chicken Little, then exhibiting preferred interpretations of the story 'on a test', and subsequently 'recognizing' in the latter, evidence of mental prowess. In other words, knowledge of what comprises learning provides an understanding of the relevance of assorted events to a singular process. The point is that this is a thoroughly folk or 'common sense' understanding. To its practitioners, it lends to their events and actions causal texture, meaningful continuity and reliability, a sense of coherence, a morally accountable order, etc. In other words, the understanding of learning allows the member to ascribe motivational relevance to activities of schooling.

This relates to discovery by analogy. For sociology, discovery in Western science must be viewed as a folk or common sense practice, or as an event whose significance as a discovery is provided for by local schemes of interpretation. The sociological understanding must bracket methodologically the naturalistic status of discovery and learning. What becomes crucial is that discovery is a motivational scheme of interpretation. In other words, events take on the character of discoveries because they appear in courses of action called 'doing research'; and conversely, that an event is defined as a discovery lends to the activities associated with it the imprint of research. The simple point here is that discoveries are discoveries because of the motivational matrices or the schemes of interpretation which our culture provides for their designation.

Common understandings and linguisticality

The position advanced here is coherent with an inspired approach to social research advocated by Peter Winch.[6] Winch, following the thrust of Wittgenstein's analysis of language, suggested that the task of epistemology and the task of sociology would merge when one inspected the way in which human cultures were embodied in language. Language, as philosophers like George Herbert Mead and the phenomenologists Husserl and Heidegger have noted, not only 'communicates information', it constitutes and forms the social relations of its speakers. In other words, when an individual learns how to speak, he/she learns the texture of the world, the way things are perceived to be connected, the way things are grouped under natural domains, and the common sense criteria that go along with and structure the 'recognition' of such events. G. E. M. Anscombe illustrates this argument with reference to the concept of causality. She argues that 'how we come by our primary knowledge of causality' is that 'in learning to speak we learned the linguistic representation and application of a host of causal concepts'.[7] In other words, verbs like scrape, push, knock over, hit, manipulate, etc., are intelligible because of the sense of causality learned simultaneously with them. Anscombe goes on to say that 'if we care to imagine languages in which no special causal concepts are represented, then no description of the use of a word in such languages will be able to present it as meaning *cause*'.[8] Simply put, the intelligibility of the world is an attribution of our conceptual makeup, and this makeup is built into our language, and becomes taken for granted and 'natural' when we are socialized into a culture. This applies not merely to the concept of causality, but to all our concepts, including the concept of discovery with which we are concerned here. According to Winch, 'there is no way of getting outside the concepts in terms of which we think of the world . . . The world *is* for us what is presented through those concepts'.[9]

Consequently, the sociologist ought to focus his attention on the meaningfulness which animates human behaviour. Human behaviour is distinctive because it is motivated or reasoned, and therefore is able to be inspected by the self and others as following some rule or standard. Consequently, learning to do anything is learning the concept which makes the behaviour meaningful. As Winch points out, only in a political society is 'putting an X on a ballot' the same as 'casting a vote';[10] who knows what sense such an activity would have in a tribal society such as Zimbabwe. One important element learned along with meaningful behaviour is the concept of a mistake or error. Winch suggests that 'the notion of following a rule is logically inseparable from the notion of *making* a mistake',[11] because in learning to do or understand something, we 'acquire the ability to apply a criterion'[12] to assess whether the thing is done correctly, or properly, or done at all.

In what follows I would like to explore the relevance of Winch's views of meaningful behaviour for understanding discovery. By viewing discovery as a meaningful act, we should be able to inspect the criteria by which it is defined, recognized, and constituted by members of society. Though criteria of intelligibility are not hard and fast, we should nonetheless be able to create a synthetic list of elements which are characteristically enforceable in the understanding of discovery. As Winch indicates, these elements are available by virtue of their use by members of society in detecting problems or errors in the ‑assessment of a concept or activity. The list I propose is synthetic in that all the elements may not be at issue at any particular time in the context of a specific discovery; rather, they have been synthesized over a series of cases. Lastly, I wish to point out that the list is neither exhaustive nor particularly surprising. Presumably, other elements could be added at some later date, when we find more of the most obvious things in the world going into the constitution or attribution of discovery.

What follows is a discussion of four criteria which I think are typically associated with the concept of 'making a discovery' and which consequently are attributed to candidate achievements when we think and speak of such events as discoveries.

THE ATTRIBUTIONAL MODEL AND THE ASSESSMENT OF DISCOVERY

The attributional model of discovery begins with the observation that the occurrence of discoveries in a culture must be viewed, not from the naturalistic question of what made them happen, but, following Winch and Wittgenstein, from the question of how they were identified as discoveries. Geographical discoveries illustrate this point well. In spite of the fact that the North America land masses were visited repeatedly by the fifth-century Phoenicians, the Irish, and the Norse as well as nomadic Siberians two hundred centuries before the voyage of Columbus, this culture gives recognition for the 'real' act of discovery to Columbus.

Similarly, when we inspect the facts surrounding the discovery of the laws of inheritance (as we shall do in the next chapter), we see that the laws of dominance, segregation, and independent assortment were not discoveries because they were cogitated by Mendel during the period 1856–70, but because in 1899 and 1900 several biologists used his experiments to address the question of Darwinian evolution and its limited ability to account for inheritance. Surely Mendel's case cannot be explained by describing how the ideas of dominant/recessive anlagen occurred to him. It is clear that the process by which ideas come into the mind takes a back seat to the issue of how ideas 'come into' the society. Clearly the explanation of discovery is tied to the latter question.

When we reflect on the question of how discoveries occur, we invariably have in mind certain achievements whose statuses as discoveries are unproblematic. We might think of Einstein and relativity, or Darwin and speciation. In other words, we begin with what are socially bona fide discoveries and then try to reconstruct the steps which led up to them! However, consider certain events whose status as discoveries experienced rather uneven receptions.

The Piltdown remains were widely accepted for at least two decades as a discovery of the missing link between *Homo sapiens* and the earliest hominids. The present interpretation of these remains waivers between charges of fraud versus hoax. However, in neither case would we expect that an examination of the cognitive development of the person behind these displays would yield anything pertinent to their social status. In point of fact, we are more liable to enlist the aid of personality psychology for such bizarre cases, preserving cognitive psychology for our more socially sanctioned specimens. In other words, how the event initially becomes formulated determines the sort of explanatory context we think appropriate. Obviously, the prominence of Lysenko would not be amenable to a cognitive explanation which described how he developed his ideas. In this case we would desire a model which focused on the socio-political conditions under which a neo-Lamarckian idea like vernalization was accorded social significance at the expense of both competing theories (Mendelian) and rival theorists (Vavilov). In this case, the social ascendance of these views appears to be a function of their ideological context: they were consonant with the dialectical materialism of the Central Committee of the Communist Party.[13]

Similarly, Piltdown appears to have gained ascendance as a result of the community expectation that just such a discovery would be made; the same with the nineteenth-century discovery of the primal form of life, *Bathybius haeckelii*.[14] This primitive amoebic form of life representing a transition from inorganic to organic organization figured as a discovery for several years, not because of the way in which it occurred to Huxley and Haeckel but because of the social recognition accorded to the event. As it turned out, these 'primitive forms' were produced by the alcohol solution used to preserve the ocean floor 'ooze' which supposedly 'contained' them. Similarly, in 1903, Blondlot's notorious *N*-rays, which followed the discovery by Roentgen of *X*-rays in 1895, became the source of numerous investigations, publications and national conferences. In spite of the fact that their existence was confirmed independently by different investigators, and measured with consistency in different ways, they were shown in 1904 to be a simple optical artifact.[15] Though these events probably involved the anomalies and gestalt switches which naturalistic theorists focus on, clearly these were only secondary in the discoveries. Their status as discoveries was a function of the perception by the community and, in the case of *Bathybius* and *N*-rays, the scientists themselves that the events of the research were *possible, motivated achievements* which were substantively *true* or *valid* and whose announcements were *unprecedented*. I would suggest that these constitute the common sense grounds or criteria of intelligibility of discovery.

Discovery as substantive possibility and motivational schema

That a discovery is 'possible' refers to the positive expectation states of a community that some such announcement will not be unexpected. As Darwin notes in his 'Historical Sketch' to *The Origin of Species*, the theory of evolution had been anticipated by numerous earlier writers;[16] indeed, as one commentator noted, it was 'in the air'.[17] Kuhn notes similarly that such expectations are given *both* by the operations of a viable paradigm of research, *and* by the appearance of anomalies which signal the need for a new theory. The expectation states may also operate in a negative way as when, for instance, the community is resistant to perceiving certain announcements as discoveries because these announcements have implications which cannot be accepted as possible or desirable. Recall the initial resistance of the leading Cambridge philosophers, including Russell and Whitehead, to Einstein's special theory of relativity; these thinkers, according to Lewis S. Feuer, because of the influence of G. E. Moore, tended to be suspicious of Einstein's theory because it *implied* a relativism in the area of ethics.[18] So too Bateson led the resistance of opinion against Kammerer's neo-Lamarckian experiments; acceptance of Mendelian genetics in Britain precluded any possibility that ontogenetic changes in the Midwife Toad could persist phylogenetically.[19] Consequently we might say that the attribution of discovery is structured in part by the perception that the announcement is coherent with existing knowledge in the field or that it resolves certain outstanding problems, as for example the discovery of Neptune by Leverrier explained the perturbations of Uranus, or the discovery of the bihelical structure of DNA by Crick and Watson explained the symmetry of chromosomal splitting and the templating of RNA from the DNA molecule.

That a discovery is 'possible' also refers to the perception that the discovery was the outcome of a course of inquiry that was substantively scientific, i.e., relevant to science. Certain achievements like the invention of Hero's steam toy and the child's toy top were not equivalent to the construction of steam engines during the Industrial Revolution nor the modern manufacture of aviation gyroscopes. Similarly the alchemist's preparation of elixirs and the astrologer's preparation of star charts do not constitute scientific achievements. Neither of these is held to be 'doing research'; and it is unlikely to be said of their production that they are making 'discoveries'. In other words, the notion that something is a discovery provides a motivational scheme of relevance in terms of which the social endeavours are framed as 'doing research'. Conversely, events take on their character as discoveries because they are seen to appear in courses of research action. This introduces the second point.

By saying of an event that 'this is a discovery', we have provided a motivational frame of reference to the event by which we exhibit understandings about the intentions of the scientist, his originality, the nature of the achievement, etc. By contrast if we formulate the event as a 'fraud', this draws on different understandings about the person, his motivation, and the ethics of his behaviour. As noted earlier, one of the problems in the naturalistic understanding of discovery is its confusion of discovery with learning. However, in the laboratory replication, the student is following a course of institutionalized inference whose disposition is known prospectively and can be assessed by comparison with the original discovery; by contrast, in the original discovery the procedures which were relevant to an outcome were often only available after certain speculations and after the event was recognized as a unique and unprecedented achievement. Clearly the events are separate motivational endeavours.

Accidental discoveries

Discovery as a motivational schema is not merely an observer's attribution or reaction to a scientific production. Though it animates the practice of scientists for observers, it also animates the scientist's understanding of his own work for himself. Characteristically, discoveries occur in the course of the scientist's in-order-to schemes of motivation as the object of his research. In other words, discovery does not 'happen onto' or 'befall' the researcher like catching cold or having a dream. These make sense only as predicates of an exogenous cause, a 'be-cause' or circumstance whose occurrence was not intentionally brought about. By contrast, discoveries are events which must be sought after and achieved.

Even so-called 'accidental discoveries', though *temporally inadvertent,* are nonetheless provided for by existing schemes of understanding without which they would be missed. Pasteur observed that fortune favours the prepared mind. However, the point here is not that the existence of a frame of relevance associated with a research plan 'sensitizes' the mind to accidental or fortuitous circumstances; rather, *it is the presence of the research as a specifically motivated course of action that makes the event accidental or fortuitous in the first place.* For example, it was research on the problem of relative densities and weight which made present to Archimedes the relevance of 'hydrostatic displacement', that is, changes in the level of bathwater. Everyone has noted such changes. That these proved decisive for Archimedes does not show that discoveries occur inadvertently and independently of motivated courses of inquiry; rather, the fact that the realization was made under circumstances ostensively motivated by other objectives lends to the latter their accidental relationship to Archimedes' scientific preoccupation. *Without this, Archimedes' consciousness of the water in his tub would neither have been an accident nor a discovery.*

There are two further aspects of the attributional view of discovery. To recognize the relevance of the category 'discovery' to an activity, it must not only be deemed a course of motivated scientific research, but the achievements at stake must be ostensively *true* or *valid*, and the announcement must be unprecedented, unique or novel.

The local validity of discovery

That a discovery is 'true' or not can be considered by sociologists as a *folk* matter. For chemists in the early eighteenth century, the phlogiston theory of matter was *'true'*, that is, it was believed to be an accurate account of the nature of matter. Likewise the primitive monera 'invented' by Haeckel and consequently 'discovered' by Huxley in ocean floor ooze seemed prima facie true, especially in light of the photographic slides that 'revealed' the primitive organization of these proto-amoebas.[20] The same may be said of the Ptolemaic world order: it persisted for a millennium and a half, only to be shown to be incorrect; yet for its adherents, it was a valid theory of the cosmos. So too the belief that dinosaurs were oversized cold-blooded lizards persisted for a century only to become contentious and problematic.[21] Similarly, the Piltdown Man stood for two decades as evidence of a form transitional between men and monkeys. In all of these cases, the perception that the discoveries were valid was a local or folk perception. Only from hindsight can we see that these theories and objects were actually invalid in comparison with our contemporary knowledge.

Conversely the relevance and validity of certain achievements in the history of science have oftentimes been missed locally. Galois' contributions to mathematics are a good example. The night before his fatal duel in 1832, the irascible Galois wrote up his theories of the solvability of equations by radicals and sent them to the Academy of Sciences, where their worth was not perceived initially because Galois' style was so obscure. These contributions were not recognized as discoveries until 1846, some fourteen years later.[22] Mendel's achievement is held in the same light; in 1865, no one – in my opinion, Mendel included –

recognized the relevance of his work to the problems of Darwinian evolutionary theory. In fact, historians argue that many of Mendel's observations had been known to breeders some decades before Mendel began his experiments in 1856. Consequently, the validity of his findings was quite a different matter when addressed to the problem of inheritance in evolutionary theory in 1900. In these cases, the recognition of validity and relevance, indeed the very status of the event as a discovery, were formulated in retrospect. The resistance to relativity at Cambridge, to Vavilov and Medvedev in Russia, and to Galileo in Rome are all cases in point. However, the interest from the sociological perspective is not whether certain communities and scientists are in fact correct (that is, after all is said and done) in their perspective of nature; rather, the question turns on whether a conclusion or theory is *perceived* as correct. In this sense, the recognition of validity is a *folk* accomplishment in the attribution of discovery. The last element to be considered pertains to precedent.

Precedence and priority

When Kuhn disregarded the relevance of Scheele to the discovery of oxygen, he did so on the grounds that Scheele did not publish in time. By publishing or making an announcement reporting the progress of his work, Scheele would not only have provided the materials that could be used to assess the *validity* of his arguments, their *pertinence* to contemporary theories and the methodological *adequacy* of his procedures, but his announcement, as an announcement, would have allowed observers to make judgements regarding the *uniqueness* of his accomplishment. That uniqueness is important for scientists and the community alike, in that it is used to warrant the originality of the contribution to the tradition, and the priority of the contributor in the community. These are separate issues. For example, as Merton points out in his study of priority disputes, Darwin was perturbed by the receipt of Wallace's unpublished paper on speciation because their positions were identical. When both scientists' views were presented at the meetings of the Linnean Society in 1858, Lyell and Hooker submitted 'for the record' that Darwin had circulated earlier statements of his position to Hooker in 1844 and to Asa Gray in 1857.[23] These statements were offered to establish that Darwin's work was original, and that he had not benefited from seeing Wallace's unpublished manuscript. But they did not establish that Darwin had priority, that he was 'first'; nor was there any dispute over this point. After all, Darwin could hardly have maintained that his earlier conjectures constituted a discovery. If they had been a discovery, how could he have justified keeping this knowledge from the benefit of other scientists for a quarter of a century? The Frenchman, E. Gley, failed in his bid to argue just such a case. In 1922, when Banting and Best announced their discovery of insulin, a medication for diabetes, Gley objected to their claim announcing that he had come to this same conclusion in 1905 and had deposited his paper in a sealed vault at the Société de Biologie de Paris. However, great exception was taken to his claim. If he had really made such a discovery, scientists asked, why had he kept this secret from the scientific community and why had he condemned a generation of diabetics who apparently went to their graves, thanks to him, all too early? Either his claim was unfounded, or he had been correct and was consequently guilty of manslaughter. Tradition has given him the benefit of the doubt and forgotten his claim.[24]

However, there have been other instances where 'sealed letters' have been prepared to secure a scientist's originality despite any evidence that rivals were on the same trail. What seemed pertinent was the scientist's sense that he might not live long enough to expound his views himself. In the case of Galois, this was obviously accurate, though his cryptic writing style scarcely facilitated the attribution of discovery.

The cases of Darwin and Columbus did not turn out so tenuously. Even before 1858, Darwin displayed a concern about his priority without knowledge of any real rivals. In 1844 after having put down an early sketch of the species question in a preliminary manuscript, Darwin wrote to his wife suggesting that should he die suddenly, she should publish the manuscript with the aid of a good editor, for the purpose of which she should devote up to £500 from his estate.[25] Fortunately, Darwin lived to oversee this project himself. Columbus responded in a similar way in 1493, when, during a violent storm on his return to Spain, fearing for his life, he recorded a version of his discoveries in a letter which he sealed in a barrel, which was then thrown into the maelstrom.[26] In these cases, the important issue is *not rival theorists* but the fear that death itself would preclude the communication of the theory. In other words, the issue is not whether the announcement of the achievement will be made *prior* to equivalent announcements by others; the issue is whether the announcement will be made *at all,* and hence whether the individual's accomplishment will be appropriately sanctioned as a discovery.

These cases suggest that Merton's analysis of priority disputes is somewhat incomplete. In Darwin's case, the concern over priority was hardly disputatious. In his first letter after receiving Wallace's manuscript in 1858, Darwin suggested that because of Wallace's presence, all his originality was smashed.[27] Clearly in his perspective it was not possible to be original if he were to be seen as 'following' Wallace. *All* his originality would be smashed because his announcement would not be seen as a discovery *at all*, if it followed on the heels of an equivalent announcement by someone else. It follows that one of the features of discovery is that its announcement changes the context in which it occurs for all subsequent, equivalent announcements. Subsequent equivalent announcements will be 'replications' or even imitations, but not discoveries. Also among competing theorists whose announcements are ostensively equivalent, it is not unusual for claims of 'fraud' and 'plagiarism' to be invoked. As Merton notes, Galileo was embroiled in 'priority disputes' with his rivals, but not over the question of who was *first* but rather over the question of who was *original*. Galileo's fury did not erupt because others were investigating the same things he was, but rather because he suspected his rivals of stealing his ideas and publishing them as their own. In other words, Galileo attributes an entirely different status to the announcements of his rivals (i.e., plagiarism): for Galileo, these announcements were not discoveries at all, and the issue was not one of who came first. Thus, while Merton's study of priority disputes points to the institutional rewards attendant on the production of scientific discovery, and the rivalry between scientists for such achievements, the present work focuses on the cognitive issue of *the very status of the discovery* in such disputes.

Hence, concern and disputes over priority often recapitulate the attributional model of discovery which has been set out in this discussion, and reinforce the conclusion that discoveries do not simply 'occur' or 'happen' naturalistically, but are socially defined and recognized productions. In other words, the question is not what makes them happen, but rather what makes them discoveries.

In summary then, this model operates in the following general way. Rather than treating discovery as a naturalistic occurrence which requires a naturalistic explanation, in this account we have examined the features of intelligibility of the phenomena of discovery which appear to ground the perception or constitution of discovery. The attribution of the status, discovery, is founded on the processes of social recognition by which the announcement of an achievement is seen to be a substantively relevant possibility, determined in the course of motivated scientific investigations or schemes of research, whose conclusion or outcome is convincingly true or valid, and whose announcement is, for all appearances, unprecedented. These are the central elements in the apprehension of scientific discoveries, both for the individual scientist and his or her community.

They are likewise the elements by which claims to discovery will be ignored or disputed; consequently, claims that an announcement is *not* 'news' or is hardly likely to be true, or is patently incorrect, will operate as a set of invalidation procedures for the disqualification of candidate achievements. Clearly, both the conditions for discovery, and for the failure of a candidate discovery, revolve around these common sense, grammatical criteria.

SEVERAL PROBLEMATIC ISSUES

Having outlined the attributional model, it is advisable at this point to address some problems or matters that might be at issue. Three items especially merit our attention.

Relativism

Any account which enjoins the reader to look on the claims of science as 'folk' claims raises immediately the spectre of relativism. After all, are not the claims of science superior to the common sense claims of any member of society, and isn't this one of the hallmarks of the scientific tradition? I have argued on the contrary that the validity of new discoveries should be examined as a sort of convention or belief. However, by arguing this, I do not mean to imply that there is, for example, no difference between science and Azande magic – for there clearly is. Though Azande beliefs and scientific theories both tend to yield self-validating interpretations, science affords far greater *control* over nature. Nonetheless, there are good *methodological* reasons for adopting a relativistic outlook. And indeed several prominent authors have already argued convincingly for it. For example, Barry Barnes suggests 'natural science should possess no special status in sociological theory, and its beliefs should cease to provide reference standards in the study of ideology or primitive thought';[28] 'Science is not a special kind of knowledge source; it has to face the problem of credibility, and the technical constraints facing the transmission of culture in any context.'[29] David Bloor argues similarly that 'the sociologist is concerned with knowledge, including scientific knowledge, purely as a natural phenomenon . . . Instead of defining it as true belief, knowledge for the sociologist is whatever men take to be knowledge.'[30]

The reason Barnes and Bloor adopt this position is that it challenges the over hasty conclusion found in the sociology of knowledge, namely, that adherence to natural science beliefs (i.e. theories) does not require any sociological explanation. Mannheim expressly left mathematics and natural science outside the purview of the sociology of knowledge, presumably because the doctrines in these areas, being objective, didn't need accounting for. Only ideology and irrational belief did. 'Nothing makes people do things that are correct but something does make, or cause, them to go wrong.' This direction in the sociology of knowledge has been countered by Bloor's study of the sociology of mathematics[31] and Barnes' examination of the 'culture of the natural sciences'.[32] These approaches have highlighted the institutional problems of socialization associated with the transmission of knowledge between successive generations and between successive traditions, and the problem of the legitimation of belief through 'metaphorical redescriptions'.[33]

My intention of looking at discoveries as folk phenomena is not inconsistent with these authors. By viewing certain discoveries in an historical perspective, we see that their meanings have changed. What is more, the progressionist character of science recommends that all our current 'orthodoxies' in science will be superseded or tempered by better theories in the future. Consequently, while at any point in time contemporary theories will be held as objective and valid, this validity has a provisional or conventional character. It is socially constructed and is likely to be superseded by later social constructions. However, as I have said, since it is historically grounded, it has a provisional validity; it is an attributed or socially constructed validity. This is what the relativistic position yields when we adopt it as a *methodological* device. However, *ontologically*, we would not want to claim that all knowledge is a function of its social context, and that everyone, the author included, is a victim of historical circumstance, and consequently that all knowledge, whether it be Azande magic, Islamic revelation, or scientific inspiration, is equally valid, or invalid. Consequently, it is imperative to distinguish between the methodological relativism of the sociology of knowledge, and the ontological relativism typically attributed to it by its critics. As Barnes and Bloor point out, before we can develop a substantively adequate sociology of scientific knowledge, we must realize that all knowledge, whether called folly or reason, is determined. Therefore, the examination of the criteria of intelligibility of discovery as 'folk criteria' does not diminish the significance of science; it merely puts its social features into vivid relief for social analysis.

Reasons and causes

Inasmuch as the present analysis draws significantly for inspiration from Winch and Wittgenstein, especially for its use of the concept of criteria of intelligibility, many readers may be led to conclude that the analysis is too 'idealistic'. That is, the analysis seems to explain away discovery by arguing that it does not occur in the mind, it merely occurs in the language. Such reactions arise from the chasm between natural language on the one hand and the actual historical process on the other. How are the two to be reconciled? It appears to me that the present account, by focusing on the ability of members of society to constitute discoveries, may appear at first to beg the question of the explanation of discovery by attributing it to human agency – something which is often seen to confound the attempt to describe formal and causal models. However, I would propose that the criteria of intelligibility constitute a set of conditions for discovery.

Numerous writers have argued that reasons, in that they cause changes in an individual's behaviour, may be considered as causes of that behaviour. 'There is no necessary incompatibility between causes and reasons as explanations of action, indeed reasons can be listed among the causes of action.'[34] For example, if you are told that smoking causes cancer, and you cease to smoke, has not this reason caused a change in your behaviour, just as mass and distance control the behaviour of Newtonian objects? You might object that the Newtonian objects are non-reflective, while humans are. However, it can be argued that it is because of their object-status that objects are subject to a class of causes called 'Newtonian' or physical. Humans, by virtue of their symbolic abilities, are subject to symbolically mediated causes, i.e. 'reasons'. By this I do not mean that 'reasons' are merely rationalizations of underlying causes, much as Durkheim argues in *Suicide,* where subjective states can be ignored in favour of the 'social facts'. The subjective states *are* social facts. Being depressed by the death of a loved one, or being angered by the rejection of others can cause one to try to take one's life. In other words, reasons held by an individual can be responsible for very discrete effects, or choices. We tend to shy away from the usage of 'cause' in descriptions of human patterns because 'cause', deriving from physical science and developed to account for the relationships governing inanimate objects, strikes us as an intuitively inappropriate way of speaking. Since the forces controlling the planets and other objects are not motivated forces, the relationships between them are 'inherent' or 'automatic', and consequently these relationships are discretely and precisely determined. However, G. E. M. Anscombe suggests that there is an historical bias in the formation of this impression. This passage is worth quoting at length.[35]

The high success of Newton's astronomy was in one way an intellectual disaster: it produced an illusion from which we tend still to suffer. This illusion was created by the circumstance that Newton's mechanics *had a good model in the solar system.* For this gave the impression that we had here an ideal of scientific explanation; whereas the truth was, it was mere obligingness on the part of the solar system, by having had so peaceful a history in recorded time, to provide such a model. For suppose that some planet had at some time erupted with such violence that its shell was propelled rocket-like out of the solar system. Such an event would not have violated Newton's laws; on the contrary, it would have illustrated them. But also it would not have been calculable as the past and future motions of the planets are presently calculated on the assumption that they can be treated as the simple 'bodies' of his mechanics, with no relevant properties but mass, position, and velocity and no forces mattering except gravity.

Let us pretend that Newton's laws were still to be accepted without qualification: no reserve in applying them in electrodynamics; no restriction to bodies travelling a good deal slower than light; and no quantum phenomena. Newton's mechanics is a deterministic system; but this does not mean that believing them commits us to determinism. We could say: Of course nothing violates those axioms or the laws of the force of gravity. But animals, for example, run about the world in all sorts of paths and no path is dictated for them by those laws, as it is for planets. Thus in relation to the solar system (apart from questions like whether in the past some planet has blown up), the laws are like the rules of an infantile card game; once the cards are dealt we turn them up in turn, and make two piles each, one red, one black; the winner has the biggest pile of red ones. So once the cards are dealt the game is determined, and from any position in it you can derive all others back to the deal and forward to win or draw. But in relation to what happens on and inside a planet the laws are, rather, like the rules of chess; the play is seldom determined, though nobody breaks the rules.

From this we may draw the conclusion that causality may well be an appropriate way of formulating human action *if* we are cognizant of all the circumstances which temper the determination of human action. It is a commonplace that a group of individuals, subjected to the same choice situation, will elect a variety of different outcomes. However, what I think is important is that, while they each may assess the situation differently, the fact of their assessment as a condition of action is consistent throughout. By this I don't mean to invoke the ghost of the 'intervening variable' which infuriated early behaviourists. The intervening variable is a regressive notion *unless* the analyst can describe the kinds of choices or reactions which an individual might elect, and hence the intentional character of the actions which result. The fact is that we are seldom confronted with actions in everyday life which lend themselves to simple formulations like the laws of motion. This is because of the extensive behavioural repertoire and innovativeness of individuals, and their ability to accommodate themselves to, and to reconstruct their social, material and mental environments. Nonetheless, in that actions are indeed responses to the life situations and histories of these environments, it makes sense to speak of such actions as determined. Yet, in that they are enmeshed in more complex systems than that of Newton, it would be a mistake to offer an *over-determined* model of action, i.e. a simply conditioned set of laws. The very inventiveness of action makes its designation as 'causal' sound awkward. Obviously this is a metaphor borrowed from Newton with much trepidation. As Anscombe suggests, it is much better to think of causality as the *derivation* of an effect or action from a prior condition, as opposed to the automatic determination of an effect from a particular cause.

Social situations do not cause, in a simple fashion, singular effects; nonetheless, effects or actions do derive from certain situations. Depression does not cause all individuals to commit suicide; yet a suicide might be caused by depression. What needs to be understood is that 'suicide' is a conventional type of response to some initial condition, though one of several possible choices in the individual's repertoire. Consequently, for the class of causes under consideration here, that is, 'reasons' ($C(R)$), it ought to be considered that $C(R)$ does not cause E, but $E(R)$ is determined by $C(R)$. Having established this, we shall return to the main question. What is the explanatory status of the criteria of intelligibility?

As noted by Bloor and Barnes, Karl Mannheim suggested that the proofs of mathematics and the natural sciences did not require causal explanation because they were valid.[36] There is a sense in which this observation is not entirely unsound. We could argue that because men judge certain beliefs in mathematics to be scientifically valid, this constitutes a reason for believing them. In other words, the objectivity perceived by the community is a condition of their adoption. This is what I have argued. However, I have also recommended that there are other features of natural scientific discoveries which are at issue in their designation: the unprecedented character of the announcement, the substantive possibility structured by the tradition, and the motivated research context. I would recommend that *these four criteria constitute a set of individually necessary, and collectively sufficient conditions of discovery.* In other words, discoveries can be said to occur as a result of the perception by members of society that each of these elements is present, and conversely, it is held that claims regarding discovery status will be disputed should any one or more of these elements be at issue. Also, it must be kept in mind that the status of these elements is that status determined by members of society, as opposed to the status which might be determined analytically by the theorist. Consequently, this theory is an explanation of discovery in that it explains how members of society produce and recognize (i.e. explain) discovery. For the member, the elements of the criteria determine the discovery (or fraud, or artifact, or whatever); for the analyst, the criteria and their use explain the member's utterances regarding new theories or announcements.

It is imperative to keep these two levels of analysis separate for the following reasons. The behaviour of individuals, in that it is purposive or goal-directed, is *teleological;* that is, actions are typically the outcome or an end-state of a motivated pattern of activity, or are the means of attaining such end-states or goals. This self-directed character of $C(R)$ perturbs many writers because it is inconsistent with causal models of natural objects. However, the problem disappears when we realize that the circularity of conditions and outcomes is a circularity of the member of society's world. In this case, the criteria *define* the event as a discovery. However, this is not a circularity of the analyst's account of the action of members of society. In other words, while the members' account may be teleological, our analysis of this behaviour is not also teleological, just as an analysis of religious behaviour is not itself religious. Simply put, the members' domain is one with its objective discoveries and their relevant features, while the analyst's domain is the criteria used by members of society to attribute and reflexively uncover those phenomena – whether these be discoveries, frauds, artifacts or whatever. The use of the criteria is teleological, circular or reflexive; however, the position which accounts for the occurrence of discoveries via the use of such criteria is causal.

The failure to maintain this distinction will result in the analyst's explanation turning into a tautology, for, it might be argued, the conditions which we specify within the criteria of intelligibility are the defining elements of discovery. Hence, the outcome is not effected by, but defined by, the initial condition. *Voilà* tautology! To avoid this we must realize that it is the use of such elements *by members of society* which constitutes discoveries, while it is our designation of this usage as members' criteria which accounts for their behaviour for us.

Focusing attention on the criteria in use by members of society, how can we be satisfied that they are in fact criteria in use, and the criteria are at least analytically separate from the products they constitute? In other words, how do we know a criterion is an independent variable? I would suggest that we have reason to believe in the independence of the criteria by virtue of three observations.

First, while the iconic picture of discovery is one where an individual grasps a new law by a kind of flash of insight, or gestalt switch, recent commentators have emphasized the temporally extended periods of assessment which precede discoveries. Kuhn noted that revolutionary discoveries are not short, simple acts like looking and seeing; they are emergent processes. Of related interest here is Steve Woolgar's investigation of the pulsar discovery by the Hewish group at Cambridge in 1967.[37] Woolgar noted like Kuhn that this discovery was not a simple act but was a judgemental matter that transpired over several months. In the pulsar work, the elements of motivation and novelty were above suspicion – that is, these scientists were clearly pursuing a programme of research that was suggested by historical developments in radio astronomy and which required new methodological technologies. Also, the prior identification of 'scintillation effects' put the Hewish group in a strong position to detect novel astrophysical emitters. The scintillation effect is the 'twinkling' or fluctuation in intensity of radio sources of exceedingly small diameter. In the summer of 1967, Hewish undertook the study of some deep space radio sources. By focusing on the scintillation effect, he hoped to be able to distinguish radio galaxies from quasars. Presumably the smaller diameter of quasars would make them far more susceptible to scintillation as the electromagnetic energy was distorted as it passed through plasma clouds in interplanetary space. The construction of an adequately sensitive radio receiver called for some 2,048 dipoles over a 4½ acre rectangle. Since scintillation is less pronounced at shorter wavelengths, the aerial was constructed to pick up wavelengths of 3.7 metres. Consequently, Hewish had constructed an instrument whose sensitivity to scintillations in radio sources made it ideal for the detection of pulsars – i.e. rapidly pulsating radio emitters.

After the first routine surveys of the sky were recorded and scrutinized, Jocelyn Bell, Hewish's assistant, began to recognize some minor 'scuff' on the records; certain very brief signals appeared to scintillate, to recur in the same location, and to be detectable at night – a time when the scintillation effect tends to be minimal. The validation of these initial records took several months of further investigation. Earlier records were re-examined for previous undetected traces. Further monitoring was undertaken, though now with an eye to detecting the possible sources of interference. It was initially thought that the effect was due to a flare star, some terrestrial radio source, or some technical artifact. Amplified recordings were made in the autumn and on November 28 these revealed a striking pulsating object – as opposed to a scintillation from a constant emitter. These were confirmed in the following days of observation. In the weeks that followed further surveys indicated the existence of more pulsars.

The point of this case for our purposes is that it illustrates that the

validation of the effect or phenomenon involved a temporally protracted series of judgements and assessments. In other words, the validity of the phenomenon was achieved through the reproduction of
the effect, the elimination of sources of error, and the identification of
further cases, etc.

The same judgemental work went into the explanation of these observations. The explanation was tied to the possibility structure of the
tradition in radio astronomy circa 1967. As Edge and Mulkay point
out,[38] 'the prevailing views about radio sources were sufficiently loose
to accommodate the existence of pulsars'. Specifically, it was assumed
that there were many different kinds of radio sources. Secondly, the
existence of small numbers of radio stars was not necessarily inconsistent with the view that most radio sources were extragalactic. And
lastly, the theoretical work on neutron stars (as well as white dwarfs
and binary systems), though largely conjectural in 1967, provided several possible mechanisms which could account for the uncanny regularity and intensity of the pulsars. Consequently, the identification of
these emitters as 'pulsars' and the models devised to explain them
were clearly circumscribed within certain historical parameters and
reflected the assessments and attributions made possible by the state
of the methodological and theoretical traditions. The point again is that
the criterion of *possibility*, like validity, is a judgemental matter, as
opposed to a simple presence/absence attribute of the phenomenon.

I might add parenthetically that the period of time often required by
the assessment of any of the criteria makes sense of earlier writers'
references to the role of anomalies and retroduction. Rather than these
being conditions of discovery, they are probably better understood as
the temporal correlates, not of thinking generally, but of the assessment required by the criteria of intelligibility.

A second indication of the independence of the criteria is the observation that sometimes the value assigned to a new finding or theory in
science changes drastically. If we assume that the original document
remains the same, we must turn our attention to the criteria of assessment brought to bear on the document. In the next chapter I will outline the drastic reformulation which Mendel's work underwent in
1900. I have already alluded to the re-assessment which attended the
cases of Piltdown, *N*-rays, *Bathybius*, etc. Clearly, when one or more
of the criteria become problematic, the event's status changes accordingly. Hence, not only are the criteria prior in time, but the events take
their status from the criteria, not vice versa. Otherwise, their status
would be unchanging; it would be inherent in the object discovered.
This does not deny the reflexivity whereby members of society locate

the referents in the objects themselves, but draws attention to the analytic observation that, in spite of the feelings of the member of society, the statuses of events do change, and they change from without.

A last point indicating the independence of the criteria is reflected, though only obliquely, in the following observation. When a discovery claim is in doubt, the sorts of ways it can be disputed are fairly discrete, and consequently I would argue that not only is discovery constituted by criteria of intelligibility, but so are discovery's alternative definitions. In other words, fraud, artifact, forgery and error also comprise our common stock of knowledge, and also operate as criteria of intelligibility drawn on to determine the status of an event. Discovery, then, may only be one of several possible attributions that an event can be given, and any particular announcement may undergo *all* of these statuses when we consider the variance of actual responses to discovery claims. As Law and French quite correctly point out, we tend to overestimate the unanimity of response to theories in science.[39] A good illustration of this is the recurrent appearance of Lamarckian doctrines in the post-Darwinian history of biology. Also, Whittaker's history of physics is notorious for its less than qualified endorsement of Einstein's achievements.[40]

Consequently, this suggests that not only are the criteria of intelligibility of discovery something existing prior to the actual conferral of such status, but they are only one of hosts of possible statuses which could be conferred on the event. This reinforces the opinion that the criteria are a tacit method in the eye of the beholder, as opposed to a simple reading of the object's factual nature.

The preceding remarks have discussed the explanatory status of the criteria of intelligibility in terms of the cause–reason controversy. I have endorsed the notion that reasons are interchangeable with causes in explanatory accounts with the reservation that causes are best understood as conditions from which other events arise, not conditions which automatically entail effects. I have described the teleological nature of accounts of discovery by members of society, and shown how the criteria from which the occurrence of such events arises are separate from those events themselves, thus demonstrating the independence and causal status of the model. The last remarks which I offer here are to characterize the manner in which such criteria of intelligibility operate in actual settings.

Context and detail

When we consider that our common stock of knowledge provides us with numerous classifications for discovery claims, and when we consider that our technical expertise allows us to sharpen our criteria to a fine point, then the attribution of discovery takes on the following interesting status. It is in fact a *method* of elucidating an event.[41] In terms of the social context, an utterance which indicates that some particular announcement is a discovery can be read on two levels. For the member this utterance can be read to indicate that some production has a notable set of attributes – the criteria we have been discussing. However, analytically we can see that because this attribution is one among many, its designation functions for the member of society as a method of scrutinizing the details associated with the event. It is a method in the sense that any particular interpretation tends to treat every new fact as a confirming element of the overall event. All are methodically collected under the auspices of the interpretation. However, should the interpretation change, should an erstwhile discovery become a notorious fraud (as in the case of Piltdown Man), every fact takes on a new relevance in terms of the new criteria bearing down on the event.

There has been a trend recently in the sociology of science toward the detailed analysis of actual scientific work settings. The object of such studies has been to gather an account of science as it is practised in all its details. While no one could in principle find fault with such 'anthropological' approaches, the researchers run the risk of suffering from 'ethnographic dazzle' whereby they are in danger of losing sight of what all the details are details of. It is my conjecture that the details of any setting are intelligible only insofar as details are details of some imputed 'context'. In other words, 'facts' are available as 'detail' only when some interpretation of an event or setting has been made which puts it in context, imputes to it a status, and expects of it certain characteristic exhibitions. When I speak of discovery's criteria as a method, I am thinking of how this method organizes its features into intelligible, discovery related facts (i.e. unprecedentedness, validity, possibility, scientifically motivated). This leads me to suggest that the investigations by social scientists of actual laboratory work or chalk-board inference or model building behaviour will wander aimlessly unless it can establish how all these activities are conceptually intelligible. By way of conclusion, I recommend that the criteria of intelligibility work as methods for members of society, focusing their attention on those features of an event which the criteria bring under scrutiny. Ergo, the concept of discovery is a member of society's method of appraising candidate announcements in the press, as well as candidate achievements in the laboratory. This point will be illustrated in later discussions.

The utility of our model can only be assessed when we examine its power in accounting for specific historical cases. In the next chapter we shall examine the case of Gregor Mendel, and his varied importance in different traditions. In the chapter following this, we shall explore several more discoveries in some detail to extend and corroborate the perspective advocated here.

6

The law valid for Pisum *and the reification of Mendel*

In the history of genetics, the case of Mendel presents one of the significant recurring problems. How could a series of outstanding experiments which were conducted over a period of years and which laid the foundation for the modern field of genetics have failed to come to the attention of the scientific community? A great deal of historical commentary has tried to explain the 'long neglect' of Mendel's work by drawing attention to such alleged obstacles as the 'forbidding mathematical approach', and the 'obscurity' of the publication, the 'low status' of the researcher, the 'prematurity' of the problem and the misinterpretation of the results. In fact, Bernard Barber uses Mendel to illustrate nearly every single condition which contributes to the resistance of scientists to new scientific discoveries; and his opinion is shared variously by such writers as Loren Eiseley, Elizabeth Gasking, Bentley Glass, Garrett Hardin, Robert Merton, Alfred Kroeber, and Gunther Stent among others.[1] Indeed, it is an almost universally shared opinion today that (a) Gregor Mendel was the original founder of the science of genetics; that (b) the announcement of his discoveries in 1866 was initially in some way confounded, distorted, misunderstood or simply lost in the information transfer system; but that (c) his contributions were subsequently brought to light simultaneously by several researchers working independently on the same sort of problem in 1900.

This chapter offers several alternative hypotheses, namely that (a) Mendel was simply not the founder of what we today call the science of genetics; that (b) he was not overlooked in 1866; and that (c) he was not re-discovered purely and simply by three later researchers working independently on the same problem in 1900. On the contrary, the status of Mendel's work changed over time with a shift in contexts from a concern with hybridizing as a key to evolution pure and simple, to a concern with hybridizing as a clue to a theory of inheritance which would complement a theory of natural selection. Hence, the status of the discovery was not a simple function of the *contents* of the paper, as much as the *contexts* in which it appeared.

Specifically it is suggested that Mendel's revival in 1900 took place in the context of a priority dispute between Correns and De Vries, and that this dispute led scientists to overlook the original intent of the earlier research. Furthermore, the revival of Mendel in England emerged in the context of a controversy between the biometricians who had adopted a model of continuous variation and evolution by the selection of individual differences, and the 'saltationists' like William Bateson who had adopted the model of evolution by the selection of 'mutations' or discontinuous variations. In the context of this controversy, Mendel's work was erroneously employed to dismiss biometrical models of inheritance and to underwrite the efforts of the mutationists. However inaccurate such positions turned out to be, Mendel's achievement emerged at a time when the problem of inheritance was an acute question in evolutionary theory, especially in light of Darwin's failing model of pangenesis. On the other hand, it appears that in his own day, Mendel's work was undertaken in the tradition of the hybridists who viewed the process of inheritance not as a *subfield* in the general theory of evolution but as itself a potential explanation of the origin of species. These facts tend to support the opinion that in 1866 Mendel's work figured as normal science in the hybridist tradition, while in 1900 the revival of Mendel's discovery of segregation constituted a relatively revolutionary achievement.[2]

This difference in status recommends Mendel's case as a telling illustration of the attributional model of discovery. While the original paper may not have changed in any of its details, the circumstances and traditions against which it appeared led readers of the paper to attribute to it different relevancies depending on the point in history at which it was brought to their attention. In the following discussion, I intend to explore this one case to show in detail the transformations in the status of Mendel's discoveries, and hence to bear out my argument that events are discoveries not by virtue of how they appear in the mind, but how they are defined in and by a cultural criterion.

THE CONTEXTS OF MENDEL'S REVIVAL

In 1900 Hugo de Vries announced the results of his experiments describing 'the law of segregation of hybrids'. That law was based on De Vries' reformulation of the Darwinian hypothesis of pangenesis presented in his *Intracellular Pangenesis* (1899). De Vries divided Darwin's hypothesis into two parts: a *material unit hypothesis* which held that qualities inherited by the organism are represented by discrete material particles in the germ cell, and a *transportation hypothesis* which held that parts of the organism throw off particles which often become incorporated into the germ cell, and result in the inheritance of changes acquired during ontogenetic experiences. However, De Vries dismissed the transportation hypothesis on the grounds that it was not empirically supported. August Weismann had as early as 1883 propounded the theory of the absolute independence of the germ plasma from the other somatic cells; and before this, Francis Galton's transfusion experiments had thrown doubt on the existence of mobile 'gemmules' and their effects on the germ plasma and inheritance. Consequently, De Vries' reformulation of Darwin's provisional hypothesis focused exclusively on the existence of discrete particles of inheritance.

De Vries undertook a programme of experimentation during the 1890s to explore the behaviour of these particles during the process of inheritance. However, his chief interest in these experiments was to determine the process by which species emerged. In De Vries' mind, the most important hereditary solution to this problem was the process of mutation. From at least 1886 when he discovered new species of evening primrose in a field near Hilversum, De Vries had entertained the hypothesis that speciation occurs through the appearance of new species by discontinuous variation of traits. De Vries concluded that the new species of primrose (*Oenothera lamarckiana*) found side by side with the traditional but markedly dissimilar form of the same species had appeared as a result of a mutation.[3] During the decade prior to his segregation paper, he conducted large numbers of hybrid experiments with over thirty species, in which he observed the 'splitting' or segregation of monohybrids in, for example, opium poppies and *Oenothera lamarckiana;* second generation hybrids characteristically reverted to a recessive character in about one-quarter of the plants. It was his conjecture during this period that evolution resulted from what he termed 'progressive mutations', i.e., mutations in which the effect of an 'active pangen' was not held in check by a 'semi-latent pangen'.[4] Different populations presumably could be described in terms of the activity or latency of the pangens and hence classified according to their mutability. Since De Vries was most interested in the origin of species, progressive mutations, i.e. mutations which had arisen *without* 'antagonistic pangens', were of far more interest to him than hybrids characterized by segregation and reversion of parental stocks. Consequently, though De Vries recognized the ratio of dominance and the principle of gametic segregation, these were of secondary interest to his *Mutation Theory*, which he published in 1901–3. Nonetheless, he published his law of the segregation of hybrids, which specified two important conclusions:[5]

1. Of the two antagonistic characteristics, the hybrid carries only one, and that in complete development. Thus in this respect the hybrid is indistinguishable from one of the two parents. There are no transitional forms.
2. In the formation of pollen and ovules, the two characteristics separate, following for the most part simple laws of probability.

Actually, De Vries wrote three articles announcing his conclusions. His first communication was published in *Comptes Rendus de l'Academie des Sciences* at Paris. Here De Vries made no mention of the fact that his laws of segregation and ratios of dominance were identical to the conclusions of a certain Gregor Mendel which were published thirty-four years earlier. However, in a more extended report which appeared in May, and which had actually been the first of the three reports to be prepared, De Vries noted that 'these two statements in their most essential points, were drawn up long ago by Mendel for a special case (peas).'[6] In a footnote he suggested that 'this important treatise is so seldom cited that I first learned of its existence after I had completed the majority of my experiments and deduced from them the statements communicated in the text'.[7] Exactly how seldom was Mendel cited?

The Mendel citations

Mendel's paper on *Pisum* was cited several times in different places before its wide acclamation in 1900. In 1869 it was quoted in Hermann Hoffmann's *Untersuchungen zur Bestimmung des Werthes von Species und Varietät*. Mendel's conclusions were summarized as follows: 'Hybrids possess the tendency in the succeeding generations to revert to the parent species.'[8] One finds a similar interpretation in the second major book to appear, *Die Pflanzenmishlinge* by W. O. Focke (1881). Focke cited Mendel's *Pisum* work fifteen times, though again he interprets Mendel as reporting the sort of thing already known from the work of earlier hybridists like Andrew Knight, specifically that hybrids tend to revert to the form of the parents and do not exhibit a fusion of characteristics as a result of the cross. Focke also mentions that 'Mendel believed that he found constant numerical proportions between the types of hybrids.'[9] Focke's reference was copied by George John Romanes and cited in a list of plant hybridists in an article entitled 'Hybridism', which appeared in the ninth edition

(1881) of the *Encyclopaedia Britannica*. There was no specific discussion of Mendel.[10] Another reference to Mendel to appear in 1881 was recorded by Benjamin Daydon Jackson in *The Guide to the Literature of Botany: Being a Clarified Selection of Botanical Works*. This reference was also lifted directly from the citation of Focke. The *Guide* listed some 6,000 references to botanical works not contained in the *Thesaurus* compiled by Pritzel, another popular reference guide, but again there was no discussion of Mendel's contribution.[11]

The next reference to Mendel's paper appeared in L. H. Bailey's 1892 article, 'Cross-breeding and hybridizing'. Bailey had not read Mendel's work, but like the others, merely lifted the reference to Mendel from Focke's 1881 book.[12]

Among the other references we find to Mendel in this period is one recorded by I. Schmalhausen in the appendix of his Master's thesis at the University of St Petersburg in 1874. Though Schmalhausen read Mendel only after having completed his research, he appears to have understood his concept of the segregation of elements in the gamete, and the resulting ratio of the characteristics in the hybrids. Also, Schmalhausen pointed out the similarity of Mendel's work to Naudin's views on segregation.[13]

The only other references to Mendel which have been uncovered to date appeared in 1872.[14] One was in a dissertation written at the University of Uppsala by Albert Blomberg, in which Mendel's *Pisum* experiments were discussed and again compared to Naudin's work on segregation. The other paper was by Anton Besnard and appeared in a paper on *Hieracium* plants published in *Flora* in 1872. We shall deal with the interpretations and comparisons of Mendel in a moment.

While De Vries suggests that he became acquainted with Mendel's work *after* the bulk of his experiments was completed, some historians have expressed reservations about this claim. Glass points out that L. H. Bailey sent a copy of his 'Cross-breeding' to De Vries in 1892, clearly a date which follows De Vries' *Intracellular Pangenesis*, but which probably does not postdate the 'bulk' of the experiments.[15] In response to H. F. Roberts' queries in 1924, De Vries suggested that he first came across Mendel's work as a result of the citation in the bibliography of Bailey's 1895 *book*, though it appears that Mendel was only mentioned in the 1902 and subsequent editions.[16] However, this confusion is easily understood when we learn that De Vries communicated to Bailey that he had come across Mendel as a result of the citation in Bailey's 1892 *paper*. These two sources could easily have been confused in De Vries' mind. However, the confusion doesn't end there, for T. J. Stomps related that in 1900 De Vries received a copy of Mendel's paper from a friend, Martinus Wilhelm Beijerinck, a professor of bacteriology at the University of Delft 'who had himself suspected the operation of something like unitary mutation in bacteria'.[17]

Certain historians like Robert Olby suspect that De Vries may have had no intention of mentioning Mendel's earlier work and that perhaps the crucial factor in his recognition of Mendel was the swift reaction by Correns and Tschermak to the *Comptes Rendus* article. Both Correns and Tschermak had independently become aware of the phenomenon of segregation and the ratios, and had read Mendel's paper during the winter of 1899–1900. However, if De Vries had learned of Mendel before 1900, and if he had planned to suppress this fact, as Olby believes, his plans were interrupted by the communications of Correns and Tschermak. Correns related to Roberts that he received a reprint of De Vries' *Comptes Rendus* article on April 21, 1900, and that he had immediately composed his own finding which he had prepared for publication by the evening of the 22nd![18] Tschermak had conducted his research for a doctorial dissertation which he had defended on January 17, 1900. Upon receipt of De Vries' reprint, he too immediately arranged the publication of his results and succeeded in having them accepted in the *Journal of Agricultural Research* in Austria. When De Vries received pre-prints for these papers, he appears to have edited the galleys of his second and third articles before they were published. Sturtevant speculates that the unusual number of printer's corrections in the longer German version of De Vries' paper are probably indicative of a difficulty in following De Vries' hasty corrections to the galleys.[19] Also, in the second French version of the paper submitted to the *Revue Générale de Botanique,* mention of Mendel's original paper and of the forthcoming contributions of Correns and Tschermak appear to be awkwardly appended to the article, as though they were an afterthought.

In his own announcement of segregation results in the *Berichte,* Correns exhibits two distinct reactions to De Vries. First of all, he frames his announcement so as to indicate that though he had lost priority in the discovery to De Vries, both had lost out to an earlier researcher, even though the initial intent of that research, and its contemporary importance, were somewhat less than identical. In other words, he neutralizes his loss in what would have otherwise been a priority dispute between him, De Vries and Tschermak. This is accomplished decisively by labelling the discovery 'Mendel's Law'. This is perhaps the single most important fact in the reification of Mendel as the founder of genetics. The action effectively undermined the priority of De Vries' claim to the discovery, and at the same time lent a decisiveness to Mendel's experiments which they would not have had, had the community not experienced Weismann's cytological conclusions and the weakness of Darwin's hypothesis of pangenesis. Correns glossed these differences in context by suggesting that Mendel had come to the same conclusions that he and De Vries had in 1900, 'as far as it was possible in 1866':[20]

I also, in my hybridization experiments with races of maize and peas, had arrived at the same result as De Vries, who had experimented with races of very different sorts of plants, among them also with two maize races. When I had found the orderly behaviour, and the explanation therefor . . . it happened in my case, as it manifestly now does with De Vries, that I held it all as being something new. *I then, however, was obliged to convince myself that the Abbott Gregor Mendel in Brünn in the sixties, through long years of and very extended experiments with peas, not only had come to the same result as De Vries and I, but that he had also exactly the same explanation, so far as it was at all possible in 1866.*

Secondly, Correns exhibits a certain amount of suspicion regarding the frankness of De Vries and his unacknowledged reliance on Mendel. He insinuated not only that De Vries had come to the same conclusions but suggests in a stroke of understatement that the names given to the terms are 'coincidentally' the same: 'This one may be called the *dominating, the other* one the *recessive* anlage. Mendel named them in this way, and by a strange coincidence, De Vries now does likewise.'[21] This, of course, was no coincidence; De Vries was familiar with Mendel as he admitted in his second two papers. However, in light of De Vries' programme of research, it is probably fair to say that De Vries sincerely believed that his own theory of the role of mutations was far more important than the observations of Mendel on the segregation of antagonistic pangens. After all, his programme pointed to the process of species formation, while Mendel's conclusions appeared to point to species preservation; one was a general theory of evolution, the other a theory of inheritance. De Vries' opinion of Mendel's contribution did not change in the ensuing years of the development of genetics. Indeed, on October 31, 1901, De Vries repeated an earlier suggestion he had made to William Bateson:[22]

I prayed you last time, please don't stop at Mendel. I am now writing the second part of my book which treats of crossing, and it becomes more and more clear to me that Mendelism is an exception to the general rule of crossing. It is no way *the* rule! It seems to hold good only in derivative cases, such as real variety-characters.

Consequently, De Vries' later failure to mention Mendel in his 1907 *Pflanzenzuchtung,* and his refusal to sign a petition calling for the construction of a memorial to Mendel in Brünn, probably ought not to be taken as evidence of his jealousy of Mendel, as Tschermak suggested, but as confirmation of his conviction that Mendel's importance was overrated. Nonetheless, Mendel's paper was republished in *Flora* in 1900 with a recommendation of its great importance. Tschermak also succeeded in having it republished in 1901 in *Ostwald's Klassiker der Exakten Wissenschaften*. When Bateson received a copy of De Vries' work, he immediately looked up the Mendel paper, and soon had it translated and published in the *Journal of the Royal Horticultural Society of London* in 1901. In England it was seized upon at once as evidence supporting the model of evolution through the selection of discontinuous variations. This was the second major step in the reification of Mendel in 1900.

Though Mendel's work is often cited in retrospect as a decisive contribution to genetics, it is clear that in England those who initially found it so decisive were the same people who were most frustrated by the account of variation and inheritance found in Darwin, and who were exploring alternative accounts, especially in the notion of discontinuous variation and mutation. One of the great ironies in the history of evolutionary theory is that although Darwin's work on *The Origin of Species* was probably instrumental in the establishment of evolutionary thinking in Victorian England, few naturalists actually subscribed to Darwin's view of the natural selection of individual differences.[23] Such variations were thought to be too easily 'swamped' by crossing.

The controversy over variation in evolutionary theory

Darwin, as well as other naturalists, believed that there were two sorts of variability: on the one hand, individual differences, and on the other, 'sports' or mutations. Darwin believed that evolution could not progress by the natural selection of sports for several reasons: their progeny were frequently infertile; they often experienced pathogenic imbalances in the internal organs or tissues ('monstrosities'); and they occurred far too infrequently. In addition, they were subject to processes of 'swamping' just as much as selected individual variations. On the other hand, individual variations were typically heritable, they did not suffer infertility, and most importantly, they occurred in large numbers in every generation, and hence gave natural selection a great pool on which to operate.

As noted earlier, one of the drawbacks of the reliance on small individual variations was that these were very easily swamped. A given variation, unless isolated from the population, would not be perpetuated in an organism's progeny because the interbreeding of the progeny with the rest of the stock in subsequent crosses would result in the decrease in the number of elements in the germ cells which controlled the new variation. Each new generation would, therefore, tend to contain less of the germ material underlying the new characteristic. As Darwin himself noted, this process, as well as natural aversion and infertility, tended to keep species quite distinct from one another.[24] However, in geographically isolated areas, such as the Galapagos Islands, different conditions of existence together with these minor variations might be able to facilitate the perpetuation of even small variations. Nonetheless, many of Darwin's supporters thought that he was confining himself unnecessarily with such a theory. T. H. Huxley, Francis Galton, William Bateson, W. K. Brooks, and even some of Darwin's antagonists like Mivart thought that evolution by this slow and gradual process was highly unlikely. Most thought that it was more probable that evolution occurred through the operation of selection on discontinuous variations, saltations or sports. Additionally, this model of selection was not subject to the charge that had been made by certain physicists, particularly W. Thomson (later Lord Kelvin), who criticized the theory because it presumed a far longer period of geological time than geophysical theories allowed for the formation of the earth. Discontinuous variations presumably would allow speciation to occur at a far quicker rate than continuous variations.[25]

Darwin was not unaware of these problems. In 1865 he proposed an idea in a brief paper written for Huxley, which integrated a mass of diverse information about reproduction in sexually and asexually reproducing populations. This was his 'provisional hypothesis of pangenesis' which appeared in extended form in 1868 in *The Variation of Plants and Animals under Domestication*. William Provine summarizes the theory as follows:[26]

Basically the theory stated that each part of the organism throws off 'free and minute atoms of their contents, that is, gemmules'. The gemmules multiply and aggregate in the reproductive apparatus, from which they are passed on to the following generations. The theory was designed so that the 'direct and indirect' influences of the 'conditions of life' might become embodied in the hereditary constitution of the organism. If an organism were affected by the environment, the affected parts would throw off changed gemmules which would be inherited, perhaps causing the offspring to vary in a similar fashion.

Thus Darwin, by giving a role to the effects of the conditions of life, guaranteed at least in his own mind the large numbers of heritable differences, with a tendency to vary in the same direction over time, both of which conditions would be required to overcome the blending problem. However, it does not appear that many found this hypothesis convincing, for the search for models of discontinuous variation continued. We have already mentioned the efforts of De Vries and his search for mutative variation. Francis Galton, Darwin's cousin, had attempted an empirical test of pangenesis by performing blood transfusions on different varieties of rabbits, but failed to find any evidence of transmission of characteristics from one animal to another, and hence concluded that there was no evidence of 'gemmules' floating in the bloodstream.

Galton's attempt to produce a more viable theory of inheritance took the following course. Though he admitted the omnipresent evidence of variability, his observations on things like human stature led him to believe that the variability always distributed itself in the long run around a fixed median average. He expressed this in his Law of Regression, which stated that the deviation of a new organism in some characteristic will be a fraction of the parental deviation from the same population norm for that characteristic. Specifically, the new organism will inherit two-thirds of the parental deviation from the norm. Hence, the population variability will be 'in a constant outgrowth at the centre', and a 'constant dying away at the margins', thereby preserving the average at a fixed point. With this situation, Galton concluded that speciation *had* to depend on sports or discontinuities in variability, for mere individual differences were controlled by a regression to the median.[27]

However, Galton's two chief students, Karl Pearson and W. F. Weldon, who were studying his new 'biometrical' techniques, thought otherwise. While Galton assumed that the population median remained constant, Pearson and Weldon suggested that *if* the exceptional offspring continued to be crossed with other exceptional offspring, a new variant would arise in which the ancestral mean for the characteristic would shift over time as the 'ancestors' came to include more organisms with the same characteristic. This could occur either through isolation of a subgroup, or the elimination of that part of the population whose characteristic was disadvantageous. Pearson and Weldon became the leading advocates of the biometrical approach and employed it to defend the original Darwinian argument regarding the utility of continuous variation.

During this same period, William Bateson was developing his own ideas about the process of deviation. His study of the small isolated lakes of the Russian steppes convinced him that variations *were not* continuous with the changing conditions of life. He discovered that though there existed a change in salinity from lake to lake, there was no corresponding consistent change in the animal forms inhabiting the lakes. In 1894, he published his voluminous *Materials for the*

Study of Variation, Treated with Especial Regard to Discontinuity in the Origin of Species.[28] Bateson gathered 886 cases of discontinuous variation and on the basis of these expounded his views on the validity of a model of discontinuous evolution. With Weldon's negative review of the book in *Nature*, a series of what were at times heated personal confrontations developed between Bateson and the biometricians which ended only with Weldon's death in 1906. This antagonism similarly resulted in a power struggle for the control of the Evolution Committee set up by Galton under the auspices of the Royal Society. Though initially dominated by the biometricians, Bateson's differences with Weldon and Pearson led them, along with Galton, to resign in January 1900. They subsequently directed their energies toward the establishment of a new journal for the advancement of biometry. *Biometrika* appeared in 1902.

In 1897, Bateson initiated a series of hybrid experiments to explore discontinuity in variation in hybrid crosses. Though he did not discover Mendel's ratios, he did have a vivid sense of what ought to be explored. In July 1899 he presented a most prescient paper entitled 'Hybridization and cross-breeding as a method of scientific investigation'. It read in part:[29]

What we first require is to know what happens when a variety is crossed with its *nearest allies*. If the result is to have scientific value, it is almost absolutely necessary that the offspring of such crossing should then be examined *statistically*. It must be recorded how many of the offspring resembled each parent and how many shewed characters intermediate between those of the parents. If the parents differ in several characters, the offspring must be examined statistically, and marshalled, as it is called, in respect of each of those characters separately.

As mentioned earlier, Bateson heard of the Mendel paper via De Vries. Given the above statement, Mendel could hardly have expected a more sympathetic reader than Bateson. However, the publication of Mendel's work did nothing to depolarize the splits in the evolutionary community. Mendel's law of segregation was bandied about by Bateson and his group as evidence of discontinuous variability and hence speciation through the process of discontinuous evolution. Consequently the initial reaction toward Mendel's paper among the biometricians was negative. Only later was it realized that a Mendelian model could account for *continuous* changes in the (phenotypic) characteristics. This emerged with the realization that certain characteristics could be controlled by more than a single factor. Even Mendel had pointed this out in his discussion of crosses with red and white flowered species of *Phaseolus*. He noted that the flowers of hybrids were not segregated discretely into either red or white but that most plants were various gradations from crimson to pale violet. Rather than treating this as a disconfirmation of segregation, Mendel suggested that 'the colour of flowers and seeds is composed of two or more totally independent colours that behave individually exactly like any other constant trait in the plant'.[30]

During the early period of the controversy, there was *some* recognition that Mendelism and Darwinism were not mutually exclusive. In 1902 G. Udny Yule published a paper in which he outlined how the multiple factor hypothesis made it possible for Mendelism to account for continuous variations and hence made it compatible with biometry and Darwinian evolution. However, so entrenched were personalities on each side of the issue that 'Yule's excellent paper had little effect upon the widening gap between the Mendelians and biometricians. Not until R. A. Fisher's first genetical paper in 1918 was there an important attempt in England to follow the lead suggested by Yule'.[31] This indicates that there was a large element of propaganda in Bateson's use of Mendel to settle his score with the biometricians, for even when the basis of a synthesis was suggested, its importance was missed.

Clearly, Bateson found Mendel's law of segregation the answer to his problem of discontinuity. However, it is not always obvious that he completely adopted Mendel's concepts. Provine points out that in his criticism of Pearson's hypothesis of homotyposis,[32] Bateson does *not* rely on Mendel's model to construct his rebuttal. While Pearson argued that sperm cells and ova were undifferentiated like cells, Mendel believed that the germ cells were differentiated. However,

Bateson did not use the criticism from Mendel's theory because he did not believe that Mendel's 'differentiating elements' were material bodies. As early as 1893, Bateson had developed a 'vibratory theory of heredity', which did not fit with a materialist view of heredity, and he maintained this theory with some misgivings to the end of his life. It even caused him to reject the chromosome theory of heredity. . . . Evidently, Bateson misunderstood or rejected what Mendel had said.[33]

When we reflect on the special value which Mendel's theory had for Bateson in the context of his dispute with the biometricians, it is less than clear that Mendel's work was simply 'revived' from dormancy. In the controversy over continuous–discontinuous variation, Mendel's paper had a relevance which was not available in 1865. Can we conclude then that Mendel was simply 'rediscovered' in 1900? Evidence suggests that Mendel's paper had a rather different valence in the context in which it was initially written.

Was Mendel rediscovered?

Mendel had read the report of his research on the effects of cross-breeding of seven different traits in successive generations of peas between 1856 and 1865 at the meetings of the Brünn Natural Science Society in February and March 1865. Weinstein suggests that the belief that Mendel was virtually unknown before 1900 can be traced to the statements of the rediscoverers in 1900.[34] Consequently, it has been widely believed that Mendel's audience in 1865 responded politely but non-comprehendingly to his work on pea hybrids. Loren Eiseley, for example, conjectured: 'Stolidly the audience had listened . . . Not one had ventured a question, not a single heartbeat had quickened . . . Not a solitary soul had understood him'.[35] A. D. Darbishire noted also that 'the publication of Mendel's paper in 1865 [*sic*] was the throwing of pearls before swine'.[36] However, recent research suggests otherwise. Reports of Mendel's two lectures indicate that he received very positive and rather accurate coverage in the local papers. In the daily *Neuigkeiten* it was reported that Professor G. Mendel gave a long lecture of interest to botanists on the results of his artificial pollination of related species by the transfer of pollen from the pollen parent to the seed parent'. The report continued in part:[37]

[Mendel] pointed out that the fertility of the cross-bred or hybrid plants was proved, but that it did not remain constant, and that the hybrids continually tended to revert to the parental forms . . . He demonstrated specimens of relevant generations, according to which characters shared in common were transmitted reciprocally, but differing characters led to the production of quite new characters . . . Particularly worthy of notice were the numerical comparisons in regard to the character difference introduced into the hybrids in their relation to the parental forms. The enthusiastic interest of the listeners showed that the subject of the lecture was appreciated, and its delivery very acceptable.

The second lecture appears to have been just as successful. The *Neuigkeiten* reported that Mendel spoke about 'the production of germ cells, fertilization, and the formation of seeds in general'.[38] Following the second lecture, G. Niessl, the secretary of the Society, added that he had observed: 'hybridization with the help of a microscope in fungi, and algae, and that further observations in this field would not only substantiate existing hypotheses but produce interesting explanations'.[39] Vitezslav Orel reports that 'minor commentaries also appeared in other German and Czech newspapers. No doubt Mendel's lectures did not remain unnoticed'.[40] Further evidence for Mendel's reputation is found in the various obituaries written following his death in 1884. On January 7, the *Brünner Zeitung* crowned a list of Mendel's achievements with the observation, 'Above all it is necessary to point out his experiments with plant hybrids'.[41] The *Tagesbote* noted similarly, 'Epoch-making were his experiments with plant hybrids'. And the January report of the Agricultural Society's journal noted, 'Downright epoch-making were his experiments with plant hybrids. What he has done and created will remain in unforgettable memory.'[42]

In light of these observations it is hard to maintain the opinion that Mendel was an obscure figure in 1865. However, it is nonetheless clear that his work did *not* evoke an international revolution in biology comparable to that which began to emerge in 1900. The 1865 lectures were submitted as one long paper which was published in the 1866 volume of the Brünn Natural Science Society. The journal, though relatively new, was mailed to 138 international addresses, two of which were in England: the Royal Society and the Linnean Society. Similarly, Mendel's seven-year correspondence with the renowned Swiss botanist, Carl Naegeli, awoke no sense in Naegeli of the evolutionary relevance of Mendel's work. Naegeli, an advocate of a true blending model,[43] probably thought Mendel's views on the segregation of material in the germ cells were erroneous, or that his results were atypical. Naegeli was himself trying to cross-breed species of *Hieracium,* a plant which appears to fertilize in cross pollination, but which is in fact self-fertilizing. Mendel's letters to Naegeli indicate that he had also begun experiments with Hieracium; he, of course, found that this plant appeared to contradict his conclusions with *Pisum*.[44]

Consequently, we begin to realize that the value of Mendel's paper was not always compelling, even for his contemporaries like Naegeli. Sir Ronald Fisher remarks on this fact in his noted paper, 'Has Mendel's work been re-discovered?'. To this question he offers the following conclusion:[45]

Each generation found in Mendel's paper only what it expected to find; in the first period, a repetition of the hybridization results commonly reported, in the second a discovery in inheritance supposedly difficult to reconcile with continuous evolution. Each generation, therefore, ignored what did not confirm its own expectations.

From this point of view, *Mendel was not really rediscovered;* presumably a rediscovery would consist in seeing that one's own views were merely *duplications* of findings recorded earlier. It is Fisher's judgement that Mendel's work, on its own terms, was a suspicious or problematic illustration of the arithmetic ratios governing the inheritance of dominant traits; according to Fisher, Mendel's ratios were far too accurate to have occurred by chance. But was this what was 'rediscovered'? Certainly, the first generation represented by Naegeli and Focke perceived the results as identical to those 'commonly reported', not only by Mendel's predecessors but by his contemporaries like Naudin. These made his work no discovery at all, merely normal science duplication or confirmation, and not all that unproblematic. The second represented by William Bateson perceived discontinuous evolution. This might have been a latent consequence of his position, but was certainly not the focal point of the study for Mendel. Fisher's position reinforces the impression that Mendelism was not revived in 1900, but constructed then for the first time. However, Fisher should not be read as supporting the conclusion that discovery is merely a 'perspective', that is, everybody saw what he expected because of his unique point of view. The point is that, as noted earlier, the processes which made Mendel's Law so important in 1900 were historically unique. Furthermore, in 1865, Mendel's conclusions were not so entirely unprecedented as we usually think.

THE NORMAL MENDEL: HIS PREDECESSORS AND CONTEMPORARIES

The present interpretation of Mendel's work contrasts that of Gasking, Glass and Barber, the main proponents of the 'long neglect' school. Gasking, for example, suggests that 'Mendel was ignored because his whole way of looking at the phenomena of inheritance was foreign to the scientific thought of his time'.[46] His 'whole way' differed in that his experiments were directed to (a) observing the inheritance of *particular* traits in hybrids and (b) observing these patterns as arithmetic proportions. Regarding the first point, Gasking notes that the thinking of the hybridists before Mendel was directed toward the specific essence of plant species and how this intermingled in cross-breeding, and reconstituted as the sex cells formed.[47] Hence, there was no appreciation of the focus on the various *individual* traits of plants and the particular form of their inheritance; hybridists were searching for whole new plant transmissions, that is, the origin of whole new species through breeding. Furthermore, according to Gasking, biometrics – the application of mathematical models to biological patterns, was unfamiliar before Mendel's time and did not achieve popularity until years later with Galton's work.

These claims are palpably misleading. Galton's first noted work, *Hereditary Genius*, appeared in 1869, only three years after Mendel's paper was published. Nothing occurred over those three years which would have particularly favoured the reception of *his* arithmetic method. As for the claim that focus on *particular* traits had not been examined by others beforehand, this too is erroneous. In 1868 Darwin published the results of his studies of hybridization and breeding in domestic species; it specifically focused on the inheritance of individual traits. And in this respect Darwin's work was far from novel. The inheritance of specific traits had always been the preoccupation of breeders and horticulturalists whose work had appeared as early as the 1790s and which proved so useful to Darwin in the presentation of his case. Therefore, not only was Mendel's work *not* out of tune with the times but the link between Mendel and his predecessors may have been much more concrete than is generally thought. This has been emphasized repeatedly in a series of articles published by Conway Zirkle, who observed that 'much of Mendelism was known before Mendel published, and we can list the earlier pertinent contributions which were probably known to him. We may even find good evidence that Mendel was familiar with the greater part of this work'.[48] Zirkle outlines the five aspects of Mendel's theory: the principle of dominance, the principle of segregation, the 3:1 ratio of dominance to recessive traits, the perpetuation of these patterns over f_{1+n} generations, and the principle of independent assortment. He then shows that work published in the *Transactions of the Horticultural Society of London* in 1799 and in 1824, and later quoted frequently in C. F. von Gaertner's classic 1849 work on hybridization, explicitly anticipated various aspects of Mendel's work. Thomas Andrew Knight, John Goss and Alexander Seton, all working with the common pea, reported observations on both the dominance of certain traits over others and the segregation of these traits in second generations of the hybrids. For example, after crossing green and white peas, Seton noted that the hybrids 'were all completely one colour or the other, none of them having an intermediate tint'.[49] Similarly, in 1826 Augustin Sageret cross-bred two sorts of melons with five different characteristics in each parent. The result was not a blending of the various traits of the offspring but the independent assortment of the traits: 'the resemblance of the hybrid to its two ascents consisted not in an intimate fusion of the diverse characters peculiar to each one but rather a distribution, equal or unequal, of the same characters'.[50] Gaertner referred to Sageret's work thirty times. After reviewing the evidence, Zirkle observes: 'We may conclude that Mendel knew of the results obtained by Knight, Sageret and Gaertner and had the work of Seton and Goss called to his attention.'[51] And what of the 3:1 ratio? 'A precise hybrid segregation ratio had been published 11 years before Mendel's paper'.[52] Its author was a fellow cleric who lived in nearby Silesia, a beekeeper by the name of Johann

Dzierzon. Dzierzon crossed German with Italian bees and found that the unmated hybrid queens produced German and Italian drones in equal numbers in a definite one-to-one ratio. His findings were published in 1856 and read in part:[53]

If [the queen] originates from a hybrid brood, it is impossible for her to produce pure drones, but she produces half Italian and half German drones, but strangely enough, not according to the type – not a half and a half intermediate type – but according to number, as if it were difficult to fuse both species into a middle race.

This work was not widely known to evolutionary biologists but was familiar to professional beekeepers. However, it would hardly be likely that Mendel, who raised and bred honey bees for two decades, would have been *unfamiliar* with Dzierzon's research. He himself kept records of the inheritance of various traits of his own hybrids. The evidence indicates that the idea that inheritance of particular traits could occur in discrete proportions might well have been suggested to Mendel by this earlier work by a fellow apiarist and cleric.

Thus on the basis of Zirkle's research it would seem difficult to argue that Mendel was ahead of his time or that his points were unorthodox compared to the existing tradition in horticulture. If anything, Mendel's reputation was modest not because he was so radically out of line with his times but because his identity with his contemporaries was so complete. His observations on segregation and independent assortment were recorded by his predecessors and the focus on inheritance ratios was pioneered by his contemporary.

Also, as pointed out by Blomberg and Schmalhausen, Mendel's results were not unlike those of Charles Naudin, the Parisian who in 1860 won the *prix des sciences physiques* awarded by the Academie des Sciences for his work on hybrids. Though he failed to describe the ratios of inheritance, Naudin clearly understood that the segregation of traits in his monohybrids resulted from the union in the sex cells of dissimilar germ material producing plants which reverted to the paternal, maternal or a mixed type. It is of note here that Darwin expressed reservations about Naudin's work for, according to Darwin, it was incapable of explaining 'distant reversion', that is, the reappearance of ancestral traits in their distant hybrid progeny.[54] Consequently, it is hard to imagine why Darwin, even if he had read Mendel, would have been any more favourably disposed towards him than Naudin.

Therefore we must conclude that the reasons offered to explain 'why Mendel's work was ignored' appear to be quite unconvincing (and, as we shall see, redundant). Mendel was clearly integrated into the tradition of hybridists. Indeed, his paper seems to begin in some debt to his predecessors and ends with a fulsome preoccupation with their questions. And as for the 'statistical' approach, this is hardly a forbidding aspect of his paper inasmuch as the great proportion of Mendel's data simply illustrate the 3:1 arithmetic proportions found in the various generations of hybrids. There is nothing especially mysterious or forbidding in this practice; indeed, if, as Fisher has suggested,[55] Mendel's results were doctored in favour of illustrating a clear 3:1 ratio, this should have only made the argument all the more forceful!

However, it would be a grave error to treat Mendel as no different from his predecessors. His work is far superior in two respects. First of all, not only had he observed evidence of segregation but he had observed the *ratios* in which the characters appeared in both hybrid and di-hybrid crosses. Secondly, he had formulated an explanation of these observations in which he attributed the ratios to the *segregation* of factors in the sex cells, and the *dominance* of certain characters over other characters. Neither achievement had occurred before. However, the question still remains as to how Mendel himself regarded his own investigation. Did he realize the value it had in the context of evolutionary theory? I would suggest that it appears from Mendel's paper that he probably did *not* appreciate the role his work could play in the theory of natural selection laid down by Darwin. Yet he did see his work as a contribution to evolutionary theory. After describing broadly his manner of research, he adds: 'this seems to be the one correct way of finally reaching a solution to a question whose significance for the evolutionary history of organic forms must not be underestimated'.[56] However, when he writes up his results, one does not find reference to the contemporary figures in the field of evolutionary theory. Had Mendel fully appreciated the significance of his contribution to the controversy raging in England, it is not improbable that he would have sent it to one of the leading British journals, as opposed to the local society, but he did not. However, this is not especially surprising when we realize that the problem of hybridization and its relationship to evolution had been a perennial theme in the Natural Science Society since its founding by Mendel and others in 1862.[57]

Even so, if Mendel was certain of the significance of his work within the framework of Darwinian evolution, it is unclear why he never sent copies of his paper to Darwin, Wallace, Huxley, Hooker, Agassi or any other proponents of the theory of evolution with whose work he would have necessarily been familiar if he were to have appreciated the subsequent value of his own contribution; but he did not. Similarly, if the value of his work was obvious, it is unclear why it did not find its way indirectly by referral and citation to the attention of those hotly debating the issues of evolution in the mid 1860s. It did not, yet it enjoyed a certain currency just the same, as a replication of commonly reported hybridization results. And this was the way in which it subsequently came to light in 1900. At that time its significance as a clue to the process of discontinuous evolution and particulate inheritance was apparently obvious above all to Bateson who quickly put it to work for him in his debate with the biometricians.

Lastly, if Mendel was really conscious of the significance of his ratios, it is unclear why this was never vividly expressed in the original paper itself. In that paper, Mendel certainly does formulate his findings as general observations stated in italics. For example, 'transitional forms were not observed in any experiment'.[58] Also, 'It becomes apparent that of the seeds formed by hybrids with one pair of differing traits, one half again develop the hybrid form while the other half yield plants that remain constant and receive the dominating and recessive character in equal shares.'[59] Hence the regularities are stated explicitly and, in fact, further on, having reviewed the experimental results for peas, he refers several times to the *lawful* character of the regularities: 'the law of development discovered for *Pisum*', 'the law of the simple combination of traits', 'the same law as in *Pisum*', and 'the law valid for *Pisum*'. However, this usage seems to be offered tangentially, that is, it does not seem to draw attention to the fact that the identification of the law is the whole point of the paper, and that such a law governs inheritance generally. In other words, *this is never explicitly stated either in or as the conclusion.* Unlike Darwin, who prefaced his treatise with a discussion of the relevance or importance of his own work to 'that mystery of mysteries', evolution, and his proposed solution, Mendel only obliquely ties his research to 'the evolutionary history of organic forms'. When Mendel refers to the larger issues of evolution and to other theorists,[60] he discusses not his *own* work, or even that of Darwin, Huxley, or Wallace, but that of C. F. von Gaertner, Koelreuter and Wichura – the hybridists who had explored hybridism as a clue to the question of speciation. If Mendel had been thinking of evolution and heredity in the way later researchers thought of these things, he is certainly less than frank in communicating this to his readers. The significance of his work to the larger issues of evolution and natural selection appears to be obscured by his discussion of the work of other hybridists.

A solution to this paradox might be offered by the observations of G. Niessl, the secretary of the Brünn Natural Science Society in the early 1860s and still an active official there at the turn of the century.

In 1902, he suggested that it was believed in the 1860s that Mendel's work was in *competition* with, as opposed to *complementary* to, that of Darwin and Wallace: 'His work was well known but ignored in the prejudice of the then exclusively *different divergent* views . . . for the principle of the then generally acknowledged hypotheses of Darwin were almost exclusively decisive.'[61] In other words, it seems that the Darwinian model (focus on selection processes) appeared inconsistent with the Mendelian model (focus on combination of traits). Also, Vitezslav Orel suggests that Mendel never used the terms 'heredity' or 'hereditary' in his analysis, and that when later evolutionists wrote about Mendel, they were assuming that his work figured under the broader umbrella of evolutionary theory, i.e. that heredity theory was a subfield in the explanation of the evolution of new species. Consequently, as Correns admits, they were attributing to Mendel a status different from what he himself pictured. As Correns noted, 'these sentences are not formulated by Mendel himself, but were derived from reality only at their rediscovery'.[62] In other words, Mendel seems to have believed, like De Vries, that an account of heredity would be *equivalent* to an account of evolution; that is, Mendel was seeking an explanation for the process of evolution which did not require a theory of selection. This reading of Mendel was apparently overlooked in 1900 when, for Bateson and company, a theory of selection was already taken for granted in the debate with the biometricians, and when the question of inheritance was the crucial missing link or subtheory that consolidated the broad model of evolution. In other words, the work of Mendel was revolutionary in the context of latter-day evolutionary theory, where it constituted a model of heredity, an evolutionary subfield; but in 1865, when viewed as an account of evolution based on a model of hybridism, it appears to have had only mixed success. To appreciate this last point, one needs to re-examine Mendel's discussion of Gaertner, Wichura and Koelreuter.

The hybridist tradition referred to by Mendel

In his introductory remarks to the 1866 paper, Mendel refers to the work of Koelreuter, Gaertner, Herbert, Lecoq, and Wichura who had 'devoted a large part of their lives' to the problem of hybridization. These hybridists had all conducted numerous experiments on the creation of hybrid plants through the artificial fertilization of stable, closely related species, and had studied the persistence of changes in the progeny. Koelreuter, in the latter half of the eighteenth century, had, in fact, produced some 500 hybrids involving some 138 different species. Gaertner in the early and mid nineteenth century conducted some 10,000 hybrid crosses with 700 species belonging to 90 different genera of plants, and obtained some 350 hybrid plants.[63] Wichura, whose memoir on the hybridization of certain species of willow plants appeared in 1865, succeeded in making some thirty-five successful crosses between twenty-one different species of willows.[64] And Herbert reported the results of numerous crosses in different species of ornamental flowering plants and in certain vegetables. These hybrid crosses were undertaken to explore the role of hybridization in the development of new species. Though the hybridists all appear to have noted the frequency of *reversion* to the grandparent species in the progeny of self-fertilized hybrids, and noted likewise the infertility or the characteristically low fertility of true breeding hybrids, it would be a mistake to conclude that they all took this as evidence of the impossibility of new forms resulting from hybridization. This is far from the case. Roberts notes that 'Koelreuter had shown that fertile hybrids could be produced between plants of different kinds'.[65] Also, Koelreuter's studies of the cross pollination of similar species by insects 'tended to cast doubt, and to require the substitution for the doctrine of the *fixity* of species . . . the principle of the *comparative* stability of organic forms'.[66] Lecoq observed in a similar vein that the process of artificial fertilization would allow the gardener the power to mix and produce species almost at will:[67]

The most difficult thing was and always is the shattering of the stability of the first type, the breaking of its habit; just as soon as an impulse thereto is present, then variation begins to know the limits of which no human eye and no human understanding suffices. With the mighty lever of hybridization in the hand, the power of the gardener is an almost unlimited one.

Herbert exhibited the most radical scepticism towards the natural unit or type concept of species. He held that:[68]

There is no substantial and natural difference between what botanists had called species and what they had termed varieties, the distinction being merely in degree, and not absolute . . . Any discrimination between species and permanent varieties of plants is artificial, capricious, and insignificant.

However, these views were based as much on his belief that species' characters could be affected by common soil, climate and the like as his belief in the power of hybridization in producing new species.

Given the preoccupation of these hybridists with species formation through cross-fertilization, Mendel's own intention to contribute to the discussion of 'the evolutionary history of organic forms' becomes more intelligible.

What evidence in the 1866 paper suggests that the role of hybridization in speciation was of central concern to Mendel? Mendel introduces his discussion of the procedure of the experiments with the observation that when varieties are crossed which have traits in common, these traits are passed to their progeny unchanged 'as numerous experiments have proven'.[69] However, pairs with differing traits 'form a new trait'. 'It is the purpose of the experiment to observe these changes for each pair of differing traits, and to deduce the law according to which they appear in successive generations.'[70] Mendel goes on to describe the traits he selected for hybridization, and to report the results for the first and subsequent generations. These results show that his hybrids were, in the first generation, either *all* like the male or *all* like the female plant, and in the second generation, again like the original male *and* the female. There were no transitional or intermediate forms like those reported by Koelreuter, Gaertner or Wichura. However, he later suggests that there is a difference between his crosses with *Pisum* which always demonstrate segregation and the reversion to parental forms, and the hybrids discussed by Gaertner and Wichura which breed true and which constitute new species:[71]

We encounter an *essential difference* in those hybrids that remain constant in their progeny and propagate like pure strains. According to Gaertner, these included the *highly fertile* hybrids *Aquilegia atropurpurea,* [etc.] . . . ; according to Wichura it includes the hybrids of willow species. This is of particular importance to the evolutionary history of plants, because constant hybrids attain the status of *new species*. The correctness of these observations is vouched for by eminent observers and cannot be doubted.

Having made this observation, Mendel goes on to describe how such stable hybrids might be organized at the level of the germ cell so as to be able to propagate purely.[72]

When a germinal cell is successfully combined with a *dissimilar* pollen cell we have to assume that some compromise takes place between those elements of both cells that cause their differences. The resulting mediating cell becomes the basis of the hybrid organism whose development must necessarily proceed in accord with a law different from that for each of the two parental types. If the compromise be considered complete, in the sense that the hybrid embryo is made up of cells of like kind in which the differences are *entirely and permanently mediated*, then a further consequence would be that the hybrid would remain as constant in its progeny as any other stable plant variety.

Clearly this passage was written by someone who appreciated the possibility of speciation via hybridization and had a vivid sense of the processes which might make this possible. Consequently it is hardly surprising that in the last four pages of the paper, Mendel discusses what would be the otherwise unintelligible transmutation experiments of Koelreuter and von Gaertner and proposes a solution to the observation that certain of these experiments took longer than others to succeed. Mendel also reported his *own* experiments on this subject. The hybridists took certain closely related species, cross-fertilized them, and continued to fertilize the progeny with pollen from the species whose re-creation was the object of the crossing. Mendel points out how the transmutation can be accounted for by his model of segregation and random combination of elements in the germ cell. He notes particularly that if there are a small number of plants and a large number of traits which are originally dissimilar, it will take longer than if the traits are quite close and there are large numbers of plants from which to choose in each successive generation. 'The transformation of widely divergent species cannot be completed before the fifth or sixth experimental year.'[73] However, the transmutation *is* possible.

Mendel concludes the paper with a reference to the fact that because von Gaertner found the transformed hybrids to remain stable, he argued against 'those scientists who contest the stability of plant species and assume continuous evolution of plant forms'.[74] Presumably, von Gaertner was arguing against people like Herbert and Lecoq who seemed to believe that artificial fertilization and the cultivation of hybrids made species *infinitely* variable and unstable. It appears that von Gaertner and presumably Mendel, who ends his paper with this discussion, find species and varieties to be *relatively* stable, but certainly not immovable, and hence certainly not incapable, through hybridization, of producing new *stable* hybrids, which, as Mendel had observed earlier, 'attain the status of new species'.

Mendel in perspective

These aspects of Mendel's paper indicate that his work was well within the tradition of the hybridists whose experiments he discussed. Mendel's paper is a brilliant formulation of the reasons for the observations of reversion and the rise of new traits. It describes the segregation of external characters in terms of the separation of trait elements in the germ cell, and their random recombination during fertilization. It also explains the patterns or ratios in the self-fertilized progeny in terms of the dominance and recessive character of the traits. Furthermore, Mendel represents these patterns in a cogent though simple mathematical way: 'If n designates the number of characteristic differences in the two parental plants, then 3^n is the number of terms in the combination series, 4^n the number of individuals that belong to the series, and 2^n the number of combinations which remain

constant'.[75] Mendel also describes the process by which new hybrids
can arise and breed true, though he criticizes the earlier hybridists for
their assumptions regarding the fluidity of species. Specifically he at-
tacks the opinion that 'through cultivation, species stability is greatly
upset or entirely shattered'.[76] Presumably he is referring to Lecoq who
suggested that the first step required to induce variation in plants is
'the shattering of their stability, and the breaking up of their habit'.[77]
This was achieved by varying the external conditions such as climate,
temperature and soil moisture. Lecoq was not alone in his belief about
the external conditions. As Roberts pointed out, 'It was the view of
Herbert that fertility in hybrids depended much upon circumstances
of climate, soil and situation.'[78] Mendel attacked this view also. 'It is
not clear why mere transportation to garden soil should have such
thorough and persistent revolution in the plant organism as its consequence.'[79] Hence, it is apparent that Mendel's work is erected on the
tradition of the hybridists, though he certainly does not take all of their
conclusions for granted. His own results with *Pisum* preclude this.
However, this does not mean that he had initiated an entirely new
science or was attacking an entirely new problem. Though the basics
of genetics are suggested by his description of the fertilization of the
germ cell with its separate elements contributed by each parent, it
appears that for Mendel this constitutes a theory of hybridization,
where hybridization constitutes a solution to the process of the evolution of organic forms.

   When read by his contemporaries like Focke, Mendel was perceived
quite correctly to be addressing the same problems as his predecessors
in the field of hybridization. However, these interpretations paid no
attention to the model of the reproductive process which Mendel in-
ferred from his hybrids. This does *not* appear to be the main point of
the paper, but appears to be one of the assumptions made in order to
make the ratios intelligible. Consequently, it is no mystery that Mendel
does not dwell on the 'genetic' model in his conclusion but discusses,
instead, the hybridists' efforts to transmute different species. By contrast, when De Vries discusses splitting or segregation, this is unmis-
takably the focal point of his discovery.

   In conclusion, it appears that Mendel was not an obscure historical figure, long neglected for three and a half decades. Nor was he entirely misread by those who were most familiar with his work. Nor
was he accurately read by those who claimed to have rediscovered his
work in 1900. In 1900 Mendel's work was read as a contribution to
the   dispute   between   Bateson   and   the   biometricians   over
continuous/discontinuous variation. Only later was the purely 'genetic'

orientation of his paper formulated. Yet in 1865, this 'genetic' orientation was a relatively minor consideration; it appears more to have been assumed than 'discovered', and even if discovered, the main point of the 1866 publication pertains to the role of hybridization in the evolutionary history of organic forms. In other words, in 1866 Mendel's research was a contribution to the model of evolution based on hybridization and the perpetuation of characters, while in 1900 it constituted a link between the phenomenon of variation and the mechanism of natural selection. This chapter has tried to outline the social forces affecting the reconstruction of Mendel in 1900, and the *in vivo* orientation of Mendel in 1866.

The perceptions of possibility in Mendel's work

In terms of the criteria of intelligibility, it is possible to specify the different statuses of Mendel's work in the following way. Because the motivational dimension of Mendel's work was never at issue (i.e. he was never suspected of plagiarism, alchemy, etc.), this element is relatively unimportant. Mendel's procedures, in that they followed the methods of natural science (controlled experiments, arithmetical groupings, etc.), were clearly not unorthodox. I suspect these methods also led people to omit questions about the validity of the distributions of dominant/recessive characters in the hybrids. However, the elements of precedence and possibility were more problematic. In 1900, Correns appears to have challenged the discovery claim of De Vries, quashing the news of the latter by attributing the discovery to a predecessor, Gregor Mendel. However, in 1866 Mendel's own precedence is unproblematic. He is linked expressly to the earlier hybridists, at least in terms of his own citations and in terms of the recollections and reflections of his contemporaries.

The crucial item, however, relates to the substantive possibility of his work. In the 1860s and 1870s he was treated by people like Blomberg and Schmalhausen as part of the hybridist tradition of Wichura, Lecoq, Herbert, Gaertner and Koelreuter. Consequently, he was seen to be clarifying and narrowing the contentions of hybridists who have attempted to evolve new species by hybridizing. In this light the paper appears as a reserved and conservative commentary on earlier work in view of his own careful experiments. On the contrary, when read in 1900 under the thrust of Bateson's work, the substantive possibility is re-structured. What he could substantively be read to mean in 1900 is decided by the controversy in British hereditary theory. In this context his experiments are read as testimony regarding the discontinuity of germ cell material, and references to the earlier hybridists and their difficulties in evolving new forms are read, not as a failure to get speciation through cross-breeding, but as a further illustration of the genetic discontinuity between related forms. Consequently, with the change in emphasis in the traditions, Mendel's work takes on a new possibility.

It should also be recalled that in the even later context of Russia's 'scientific socialism', with its inherent commitment to historical progress, Mendelism again took on a different possibility structure. As noted in chapter 5, the historical materialist creed of the Communist Party favoured a Lamarckian model of inheritance with continuously variable germ material which could be affected directly by the conditions of life. In this context, Lysenko's 'vernalized' wheat, for a period of about two decades, was seen to be evidence of a more realistic possibility. It was only with the demise of Stalin and the fall in status of Lysenko that the tide in heredity shifted back to Mendel. And, parenthetically, it was only during the liberalization of Czech society under Dubček that the old Mendel memorials were restored along with the Brünn monastery and library. The *Folia Mendeliana* dates from this period, first appearing in 1966. This publication, and especially the work of Vitezslav Orel, have been invaluable sources in the reconstruction of the early recognition of Mendel in the nineteenth century.

CONCLUSIONS AND CONSEQUENCES

For evolutionists who were searching at the turn of the century for some solution to the limitations of Darwinism, Mendel's earlier research was decisive in demonstrating that inheritance was *particulate*. This, however, could have just as easily been illustrated from the even earlier work of Goss and Seton, Andrew Knight, and Augustin Sageret. Also, though less decisively, the fact that the particulate traits were governed by specific ratios could have been thematized by Dzierzon who was himself anticipated by François Huber in 1791.[80] The point here is not that the earlier workers 'anticipated' those who came afterwards, but that, besides those offered by Mendel, certain earlier research provided many resources which could also be read in 1900 as clues to an adequate theory of inheritance. Had Correns come across the 1824 or 1826 issues of the *Transactions of the Horticultural Society of London,* the laws of inheritance might have been christened 'Goss and Seton's Law' or 'Sageret's Law'. However, these laws, like Mendel's, would not have been the revolutionary discoveries they came to be without the thematic relevance of second generation evolutionism led by De Vries and Bateson.

There are certain other misconceptions associated with the Mendel legend which also ought to be examined at this time. Not only have students of science typically subscribed to a distorted picture of the history of genetics, but this picture has become 'loaded' in certain specific ways. It is 'loaded' in that the normative interpretation of Mendel underwrites the objectives of certain other theories, not merely of the history of genetics, but of the conditions of change in science. Specifically, the typical interpretation of Mendel is strategic (a) for those who advocate the relevance of the Reichenbach distinction; (b) for those who assume the ubiquity of multiple discoveries in science; and (c) for those who advocate the zeitgeist argument which suggests that the *fame* achieved by good ideas is determined by their social settings. Each position hangs its cap on the typical vision of the long neglected Mendel, and each has become precarious in light of the present investigation. We shall explore these three issues for the implications they raise for our own views on this matter.

Reichenbach's distinction and injunctive utterances

The first major theme concerns Reichenbach's distinction between the context of discovery and the context of justification. In the case of Mendel, this distinction may not be as useful now as philosophers once thought. Discovery is not initially a psychological phenomenon which comes into the mind of a scientist, only later to be accorded a measure of recognition in a different context. Though there has been a temptation to read the work of 1866 as a context of discovery, and the announcements of 1900 as a context of justification, these two episodes were altogether different kinds of achievements for the parties involved, not separate parts of the same episode as has been suggested. The assumption of an object constancy for discovery in each context has been erroneous.[81]

Similarly, when we consider the different statuses which the research had over time, it becomes obvious that discovery is not, as Popper suggested, a psychological problem amenable to the same laws of explanation which would cover artistic creations. Clearly, *what* that discovery consisted in would vary with the tradition with which it was compared, and hence any account which *reduced* the achievement to its internal or mentalistic foundations would leave out what made the insight into a discovery in the first place.

Concomitant with the criteria of intelligibility of discovery is the subject of their annunciation in the community. Having apprehended the significance of an accomplishment, its subsequent announcement is no simple matter of reporting the information to others – as though such information had an object constancy and independent, *a priori* identity. On the contrary, the present inquiry recommends examining such announcements as quasi-performative utterances.[82] Specifically, to say 'I discovered something' does not merely document an item of information about some phenomenon, but exhibits a stance or commitment to the relevance of the achievement, the possible ways in which it could be done, its factual validity, and a confidence in the fact that this deed was not preceded by any identical achievements. In other words, the announcements or submissions of discoveries constitute *injunctive utterances*[83] which, far from simply relating the outcome of research, formulate the outcome in the context of judgements about the relevance, precedence and validity of the work reported on. As Crick and Watson state in the concluding lines of their first announcement, 'It has not escaped our notice that the specific pairing we have postulated immediately suggests a possible copying mechanism for the genetic material.'[84] DNA was not only a complex molecule, but was one whose self-copying structure had enormous importance for the biological community.

This conception of scientific writing may go some way in explaining why it is that, though certain scientists may exhibit a commitment to the same theory, there may be *no definitive statement or consensus* about all the formal propositions of the theory, their derivations and postulates, and the domain of facts which are circumscribed by it. This is because the discovery of the theory is not a discrete event whose identity is invariant, but a performative utterance which describes a world injunctively: 'Things look this way' becomes 'Look at things this way!'. Consequently, scientists who subscribe to the same theory may not share exactly the same substantial conceptions of all it entails and contains, but as Kuhn suggests may be operating within a 'disciplinary matrix' whose theories and concepts work as methods of communicating about a range of related topics, the developing prospects for investigation, and glossing without trepidation areas not accounted for by the models, images and metaphors.[85]

When we reflect on the injunctive or rhetorical aspects of scientific theories and their annunciations, the sense that some particular concrete and invariant thing has come into existence during a discovery seems less than satisfactory. This, however, is what appears to be assumed by Reichenbach in his famous distinction. An interesting case in point is the fact that in 1900 Correns described Mendel's conclusions as 'Mendel's Law', presumably referring only to the ratio of dominant to recessive characteristics. However, we observed earlier that Zirkle identified *five* key elements in the Mendelian account. These included, among other things, 'the principle of independent assortment'. Lindley Darden has shown that neither Mendel, nor De Vries, identified dihybridism (or independent assortment) as a separate law, as has become the usual practice in the meantime.[86] Even in Mendel's paper, the major propositions were not unambiguously available as discrete laws. Also, where Zirkle identifies five components of the theory, Bateson referred to Mendel's 'three primary necessities'.[87] These differences reflect the injunctive and provisional character of any particular representations of the theory.

Furthermore, though we tend to picture Mendel's revival as a universal community endorsement, this did not happen. Indeed, after the initial hiatus, the supporters of Mendel found that their views were excluded from the prominent pages of *Nature* and *Biometrika*.[88] As mentioned earlier, there was some resistance of opinion to the new theory among the followers of Galton, who had pioneered biometrics, and of Pearson, who had refined Galton's methods. Consequently, the process of Mendel's revival in Britain and America was lengthier than is usually assumed, and the career of Mendelism in Russia continued to flounder for half a century. These facts also reinforce the attributional theory of discovery outlined here. Had the discovery, or the rediscovery, and its injunctive annunciations been an open and shut matter, there would have been no resistance of opinion to Mendel. The career of Mendel's theory, first as a normal, then as a revolutionary theory reflects the processes of judgement, evaluation and attribution between the injunctive claims regarding Mendel's research and the subsequent acquiescence of opinion.

It was pointed out in this discussion that Mendel's story figured prominently in three major arguments in the science literature. We have discussed the distinction between the context of discovery and the context of justification, and themes which have emerged from this. Any recommendation of the saliency of such a distinction on the basis of the Mendel case is clearly misguided. Such a distinction erroneously assumes an object constancy for discovery which is unfounded, and ignores the attributional status of the event for both the discoverer and the community, and the quasi-performative relationship of the discovery to its annunciation as a concept or theory in the tradition.

Multiple rediscovery and the zeitgeist

The second major issue in which Mendel's case has been so pivotal is the record of multiple, independent, simultaneous discoveries. Insofar as this record is recommended on the strength of the story of Mendel, it is less than trustworthy. As we have seen, there is a world of difference between the relevance of the achievement of Mendel in 1866, and the research of the geneticists in 1900. As in the first issue, the Reichenbach distinction, the record of multiple discoveries erroneously assumes an object constancy for the laws of inheritance which is not borne out historically.

Not only is there a world of difference between Mendel and his successors, but an examination of the work of the three purported rediscoverers reveals enormous differences in their own contributions. If one consults different histories of biology, one will find that credit for the rediscovery of Mendel's work on inheritance receives inconsistent treatment. In many lists, Tschermak is omitted. Tschermak reported, after the announcements of De Vries and Correns, that he too had made the same observation as his contemporaries De Vries and Correns, but that he had done so 'phenomenally'. In other words, though he had noted the famous 3:1 ratio of dominance, its thematic relevance to evolution had not occurred to him until he had read the publications of De Vries and Correns. One might say that these announcements produced a 'gestalt shift' in Tschermak's perceptions, for he too emerged to claim a part of the priority in the rediscovery of the laws of inheritance in the summer of 1900, and supported the re-publication of Mendel's original paper in *Ostwald's Klassiker der Exakten Wissenschaften* in 1901. However, since his apprehension of Mendelism was less than self-apparent, certain authors have contested his right to the claim to the rediscovery.[89] On the other hand, if we overlook the thematic relevance of Tschermak's own research, and admit the equivalence of his work to that of De Vries and Correns, then it seems not improbable that an adequate account of the multiple rediscovery should cite not only Mendel, but his predecessors Seton and Goss, Andrew Knight, and Augustin Sageret, as well as Dzierzon and his predecessor François Huber, as co-discoverers of the laws of inheritance. Consequently, one can see how a failure to observe the differences in relevance of these various contributions would compound any record of multiple discoveries and would unavoidably confuse their significance. This is presently done in the famous list composed by Ogburn and Thomas.

There has been no detailed examination of the record of multiple discoveries. The present study of Mendel would seem to indicate that though many studies are cited as equivalent announcements on such lists, these cases may not be as straightforward as once thought. This tends to challenge the culturological belief that innovations occur with the maturation of a culture.

The third major theme in which Mendel's case has been so important is the *Weltanschauung* argument regarding the incompatibility of Mendel's ideas with the *Zeitgeist* in which he lived. Our analysis of Gasking has shown that this is incorrect. Rather than treating the absence of Mendel's reputation in the wider community before 1900 as a failure, the present argument has recommended that Kuhn's distinction between normal and revolutionary discovery, conceived as a function of the attributional status of discovery, explains the different values assigned to his achievement over time.

However, we should not overlook a crucial element behind the *Weltanschauung* thinking which, despite its methodological frailty in the eyes of the philosopher of science, explains why such thinking has such wide attraction to members of society. When Kroeber and others characterize Mendel's case as the story of monumental achievement which went unrewarded, there emerges a sense of personal tragedy in the history of science that is highlighted in the imagination because it is so clearly laden with moral issues. In science, Mendel's tragedy is rivalled only by that of Galileo. In both cases, the ordinary members of society exhibit indignation over the failure of the community to give to each man the recognition to which he was entitled. In other words, the oppression of Galileo by the Church, and the apparent obscurity of Mendel, solicit a common reaction of indignation over the apparent injustice in these cases. The same common sense of injustice animates the general interest in such cases of scientific deviance as Kammerer's Midwife Toad,[90] the Piltdown scandal,[91] and the 'suppression' of Immanuel Velikovsky. All these cases, like Mendel's, are of inordinate interest to the scientist and the layman alike in that they raise for science the moral concerns of fair play and justice which dominate everyday life.

Perhaps the moral overtones of such cases explain why Mendel's story has figured so importantly in so many positions for so long, in spite of *the lack of evidence* that for Mendel 'the law valid for *Pisum*' was a revolutionary contribution to the theory of evolution.

7

Perspective, reflexivity and the apparent objectivity of discovery

Students of science typically show a great reluctance to treat the achievements of explorers under the rubric of discovery. This hesitation would seem to be based on the fact that the great explorers were not on the face of it scientific researchers, and consequently their contributions were not, strictly speaking, contributions to science. I would suggest on the contrary that the endeavours of men like Columbus, Magellan, and Vespucci were scientific in that they revolutionized the classical cartography of Ptolemy by providing descriptions of entire new continents in the western latitudes thought to have been occupied by the Asiatic land masses. Ptolemy's *Geography,* known from a Latin translation from about the year 1406,[1] was the only authority on world geography in the fifteenth century. It calculated the distance from the most western part of Europe to the most easterly part of India to be of the order of 177 degrees of longitude. Although Columbus' estimate was somewhat greater, he was nonetheless working with a reasonably well informed hypothesis about the possibility of circumnavigating the oceans to India when he discovered, at the predicted distance, new and separate land masses: the West Indies.

It seems analytically that there is a striking parallel between Herschel who discovered Uranus while mapping a section of the astronomical sky which was assumed to be familiar beforehand,[2] and Columbus who discovered the West Indies in an ocean thought to have been empty between Europe and India. Certainly their methods of research differed extremely – the one using a telescope and existing star sketches, the other using astrolabe and navigational records and calculations. Yet there can be no grounds for counting the one land mass as a discovery because it is in outer space, and ignoring the other land mass because it is on the earth.

We have a propensity to give little credit to the work of the explorers because of the levity traditionally associated with their achievements. Recall the grammar school myth of Columbus which suggests that his voyage was initiated to prove that the world was round.[3] Far from being concerned about the spherical shape of the earth, Columbus was more preoccupied with determining exactly the distances which the voyage would cover; to this end he corroborated his predictions with the opinions of other navigators and scientists, most notably Torcanelli. When we consider that all this predated the work of Copernicus by two decades and the life of Galileo by over a century, these were no mean scientific feats. In fact Columbus' method was truly experimental in the early sense, based literally on 'experience'.

The main point here is not that geographic discoveries are scientific. Even if sound arguments could be made to show their separateness, they are nonetheless illustrative of the social basis of discovery. The issue for our purposes is raised in the question, 'Who discovered America?' The problem is nicely formulated by Hans Selye:

Was America discovered by the Indians who were here from time immemorial, by the Norsemen who came in the tenth century, or by Christopher Columbus who came in 1492? Is it still being discovered now, every day, by anyone who drills a new well . . . Discovery is always a matter of viewpoint and degree. Whenever we single out an individual as the discoverer of anything, we merely mean that for us he discovered it more than anyone else.[4]

IS DISCOVERY A VIEWPOINT?

Selye suggests that discovery is a matter of perspective (e.g. 'viewpoint', 'for us'). This suggestion is valuable as a sensitizing device, but is strictly speaking, as will be shown below, erroneous. In one sense, the same thing, 'America' (or more exactly, the Americas, north, south and central), was not discovered. The existence of the various accounts about the discoveries of parts of the New World is an artifact of the pluralism of our culture in treating the diverse explorations of the western hemisphere as though they constituted a single tradition. In other words, the various accounts are retrospective reconstructions in light of the European experience of the fifteenth and sixteenth centuries.

It is true that the *Navigatio* of the Irish monks and the *Vinland Sagas* of the Norsemen each accredited their respective heroes with the discovery. But does the existence of these records mean that the recognition of discoveries is just a matter of perspective? Were the peasants of the middle ages who held a belief in the flatness of the earth holding a 'viewpoint'? Or were they not merely holding to what they were compelled to believe was the case? It seems more plausible to hold that their perception of events was culturally bound – and similarly with the Norse, the Irish and the Phoenicians. But this is not 'a point of view'; in a sense, they had no choice in the matter in that their beliefs, though they were folk beliefs, were apparently factual, compelling and natural. Our own folk beliefs are of the same order.

From the outlook of the modern tradition, the achievements of the various earlier explorers constitute serial discoveries each of which, from retrospect, approximated the decisive one. Nonetheless it is recognized that each achievement was a discovery in its own context. Furthermore, each achievement was true. That is, for the neolithic Siberians *there was* an Aleutian land bridge. For the pre-Christian Phoenicians who explored the St Lawrence, *there was* a great river and a new continent. For the Chinese Hoei-Shen *there was* a pre-Mayan Mexico in the sixth century A.D. For the Irish monks *there was* an Iceland. For the Norse *there was* a Greenland and a Vinland. For Columbus *there was* the Indies. Each recorded an account which was for each culture factually correct; the fact that the announcement or record was in a social context was never locally at issue. Even in retrospect the social context of the recording does not *ipso facto* make it perspectival.

The problem here appears to originate from the chief premise of mundane reasoning in our culture:[5] that the world is a singular natural world which we know in common. Therefore suggestions which indicate that the world is one place for you and quite another for me are inherently problematic. Specifically the idea that an event could be a discovery for one person and something else entirely for another plainly contradicts the reciprocity of perspectives.[6]

Melvin Pollner has outlined sets of procedures which are brought into use when such problems arise. For example, in traffic courts the basic thesis that the world is a world known-in-common routinely comes under duress. A policeman presents one account of an unlawful incident; the plaintiff counters with an inconsistent rebuttal. Such disjunctures are routinely dismissed by allowing for perceptual limitations or bias, by re-interpreting the times and places which were referred to, by impugning the motivation of one of the parties, etc. In each case, the basic thesis of mundane reasoning is preserved at the expense of the disjuncture. Drawing from the classical work of Evans-Pritchard, Pollner shows that the same patterns of thought-maintenance function in Azande magic to preserve the infallibility of the poison oracle; if contradictions become apparent to the Azande, these are handled by making allowances for the weakness of the poison, the interference of ghosts, the breach of a taboo, the misperformance of the magic, etc. All such accounts preserve the initial belief that the oracle cannot fail. Pollner argues that the infallibility of mundane reasoning and Azande magic are maintained in parallel ways. Both are folk beliefs in their respective societies and are employed to give a sense of order to the world. This is the background to Selye's remark.

The suggestion that discoveries have been made which are not shared in common is potentially a contradiction of the basic thesis of mundane reasoning. Currently we have seen that the evidence indicates that America was discovered even before the Europeans emerged as nation states; the Phoenicians in particular were possibly among the earliest explorers to reach 'America'. On the other hand we are well aware of the priority given to Columbus in the folkways of our own tradition. How do we solve the problem? We chalk up the inconsistency to 'viewpoint'. Fisher acted in the same way when he suggested that each generation saw in Mendel what it expected to see.

By suggesting that the various different parties claiming the discovery of America constitute matters of viewpoint, Selye is leading us to believe that the difference between the various accounts reflects, not a difference in the structure of the world, but a difference between the people in it. We know that certain things like political opinions can be said to differ – thereby representing the personal opinions, beliefs or temperaments of voters. We speak of such parties as having their own points of view. The things designated typically as matters of perspective or viewpoint are in fact *differences*. Opinions, thoughts, feelings, etc. can *differ* without this undermining the natural assumption that we all live in the same world. This is the force of Selye's reference to 'viewpoint'; viewpoint reflects *the viewer. The world* is thereby saved from disjuncture.

However effective this argument might be within and as part of our folkways, it is not adequate as a *sociological* account. If the identity of discoveries were merely a matter of *viewpoint*, if discoveries were what just happened to be called or labelled discoveries, we would be struck by the fact that such statuses were purely arbitrary and *relativistic*.[7] While we may as analysts be prepared to accept this, we do *not* see members of society *casually* reflecting on whether they are going to hold the opinion that something is a discovery or not. As Pollner notes, mundane reasoning has a *compelling* and *natural* character for its practitioners.[8] Garfinkel notes that adherence to even the *routine* idioms of interaction, in spite of the fact that they may be relative from society to society, are binding on the members of particular societies as *moral* orders where one's identity as a competent social member is assessed and sanctioned.[9] Similarly discoveries can hardly be taken indifferently by members. As the priority disputes in science point out, the claim of discovery is neither taken nor admitted lightly. We have discussed how certain very specific features are enforceable and accountable in such discoveries; these are all grounds on which the candidacy of discoveries can be contested, and on which adherence can be enforced.

One of the inherent problems in an investigation of this type is that we are bereft of easy access to common sense reasoning as a *phenomenon* precisely because it is our own idiom.[10] Specifically, the idea that events figure as discoveries in that they are news, true, possible, etc. is hard to treat seriously as a folk idiom precisely because *for us* this is what a discovery is! The fact that the same understandings may be in force for others elsewhere with the opposite result appears to make discoveries relativistic or arbitrary. This is what Selye's view reflects. However, the proper object for a sociological study of science is how such discoveries are in force *locally* in such settings; while the sense of inconsistency or contradiction is a phenomenon of folk understanding, this does not make it a premise of sociological understanding. Though methodologically valuable, such observations are not the main focus for sociology. The focus is the apparent objectivity of discoveries for those who make and recognize them.

Discovery as a social scheme in folkways

Speaking concretely, we are more likely to accredit Columbus with the discovery because we belong to the civilization that was made present by his disclosure of the New World. In other words, we are more tied to the announcement made by Columbus because of its outcome and the structure of recognition given to it, than to the oral sagas of the Irish and Norse. We belong to the tradition that was initiated by his achievement; hence for us, Columbus' primary placement *vis à vis*, say, the Norse is common sensical. The present tradition did not depend simply on 'the New World' having existed. The fact that this hemisphere became 'the *New* World' is a reflection of the historical horizon of the European world of which we are a part. In other words, our tradition is temporally organized around Columbus' announcement of these new locations in a merchant Europe which was contextually apprised of the significance of its existence. It can be argued that it was only in the aftermath of, and directly as a result of the voyages of Columbus in the West Indies that 'America' became formulated as a reality, a real, exploitable, settlable, and discoverable locale (i.e. Spanish 'realty'). Hence it is only in retrospect that the earlier voyages were constituted as voyages to 'America'. In this sense discovery is not an atemporal *per*spective, but is temporally retrospective and prospective in terms of the actual courses of action in which discovery is conducted.

THE DISCOVERY OF AMERICA AS COURSES
OF ACTIONS

Morison described the discovery of America as 'the greatest serendipity in history'.[11] Perhaps the exploration was a happy coincidence on the basis of the fact that it was just *those* islands which were encountered, especially Guanahani.[12] However, the overall organization of the voyage was rooted in several decades of preparation. Columbus was not especially equipped to discover Guanahani in particular, but was prepared to reach any land mass or island – presumably off the coast of India. It was in anticipation of meeting some such natives that a sizeable cargo of trinkets – hawk bells, buttons and beads – was carried. This anticipation was part of the prospective structure of the exploration. In like vein the crew was provisioned with stores of food, brandy, wine, beer and water sufficient to sustain not an indeterminant voyage, but a one-way voyage of some 4,000 miles.[13]

Columbus' conjectures about the existence of lands on the western horizons of the Atlantic were initiated by repeated finds of 'big canes, pine tree trunks and wrought wood' in the course of numerous trips in the eastern Atlantic off Africa. It was during this period that he prepared his first plans for a westward voyage. Failing to raise financial backing from private enterprise, he sought royal support. When the crown of Portugal turned down his proposal he approached the Spanish crown. His request in Spain was turned over to a commission headed by Hernando de Talavera which reviewed the idea for four years and subsequently advised Ferdinand and Isabella against supporting the venture. It seems that the only reason the expedition got together in 1492 was that Columbus announced that he was seeking the support of other royal sources in Europe. Fearing that any possible benefit from such a gamble might end up in the coffers of its rivals, the Spanish crown capitulated.

Consequently, it might be argued that what was actually come upon in the voyage was serendipitously found; nonetheless its observation was set in a temporal course of action initiated over two decades, and it occurred as an event whose candidacy as a discovery was assured by the self awareness of Columbus, his captains, the crown, and the seamen that what they were doing was exploring uncharted parts of the world. Furthermore, the fact that the backing was royal lent to the research a political-legal dimension by which the geographic discoveries counted also as territorial conquests entailing juridical implications. All this assured to possible discoveries a background against which they figured as an institutionally objectified set of events.[14]

The routine activity on the boat itself was similarly structured by the motive of discovery. Seamen were constantly on watch for signs of land. Sightings of seaweed, whales, debris, pelicans and other sea fowl were duly reported to the captain as prima facie evidence of the presence of land; and the direction of the flight of birds was taken as evidence of where land lay. Also, the crown had posted a reward of 10,000 maravedis to be paid annually to the first seaman to sight new land; needless to say, so eager were the seamen that a number of unconfirmed sightings were reported before the sighting of Guanahani two hours after midnight on October 12.

En route for Spain in February 1493, Columbus encountered some of the roughest weather ever experienced in the eastern Atlantic. Fearing that his ship would be lost, and with it, all records of the achievement, Columbus wrote out two identical accounts of his travels, sealed them in boxes, which he then sealed in two wooden casks and threw overboard into the sea. Each cask contained a letter advising the finder to turn it over to the Spanish crown to collect a reward of 1,000 ducats. This constituted an attempt to guarantee institutional recognition of his discoveries, even though Columbus feared that he and his men might never reach shore themselves.[15]

When he arrived in Spain he immediately wrote a letter to the court describing his discoveries. This is believed to have been identical to the letter he had earlier cast out at sea. When it was received this letter was copied, typeset, reprinted and circulated immediately among the foreign ambassadors at the court who subsequently sent the news to their homelands. It was translated into Latin in April, and retranslated into Italian in June. The letter was in such demand that in the space of ten or twelve months it went through thirteen editions and appeared in five languages.[16] No learned quarters in all Europe were not informed of the discoveries.

The achievement was accorded its official recognition when the Spanish monarchs approached the Vatican to certify Spain's title to the new lands. A series of four papal bulls emerged in 1493 giving Spain sovereignty over the discovered islands, plus all the other islands and lands south and west of them. With respect to Portugal's sea rights in the eastern Atlantic, a longitude was drawn 100 leagues west of the Azores to demarcate the western limits of Portugal's sovereignty. And so a discovery which was conceived over a number of years off the Ivory Coast, formulated concretely in Lisbon, financed in Barcelona, realized in Guanahani, and recorded for posterity on the high seas was finally institutionalized for Christian civilization in Rome.

When we say that Columbus discovered America (as opposed to St

Brendan, Hoei-Shen or Leif the Lucky) we are 'exhibiting'[17] a structure of recognition which was accorded to his achievement by the European public in civilized Europe, by various other heads of state, by the seamen themselves – including Columbus – as well as by those who had made the event a material possibility through financing, i.e. the Spanish monarchs and their merchant backers. All this points to the primacy of the structure of recognition, that is, the *social,* as opposed to the *mentalistic,* basis of discovery.[18] Had Columbus' ships been lost at sea, and his sealed notes along with them, we would have had no discovery to reconstruct.

However, we have not completed our investigation here. While Columbus did reach unknown territory, and while this subsequently came under Spanish dominion, it must be recalled that these lands were called 'The Indies' because they were thought to constitute the Atlantic coastal waters of the Asian continent. Indeed Columbus always believed that his explorations were in the neighbourhood of Marco Polo's Cipangu (Japan) and Cathay; consequently, the cartographers at the turn of the fifteenth century represented the southern coast of Cuba as part of the Asian land mass, and the coasts of Colombia and South America as a southern extension of the Asian continent. Also, the papal bull of May 3, 1493 designated the new territories as 'islands and firm land' located in 'the western parts of the Ocean Sea, towards the Indies'. However, there was some scepticism as to whether these lands were actually Asian. Columbus was enjoined to prove that in fact he had pioneered a western passage to the Far East.

The later voyages undertaken by Columbus and a host of other explorers between 1493 and 1502 sought the proof which was required. First, Columbus tried to show the continuity of his new lands with the Asiatic land masses explored by Marco Polo, and secondly, he sought the western passage through which Polo had returned from his earlier explorations. Columbus convinced himself that the coast of Cuba was so extensive that it had to be the southern part of the Chinese province of Mangi. He also established that to the south of the Indies lay 'firm lands' of sizeable proportions. Rather than treating these as evidence of a separate unknown continent, he suggested that these lands formed part of the fabled 'terrestial paradise', in an unfamiliar extension of the Asiatic land masses. Lastly, in his final voyage he found the isthmus of Panama which connected his recently discovered *mondo novo* to the south with what he took to be the traditional northern provinces of Cathay to the north. From the natives he reports learning of a thriving community of traders living on an expansive sea only ten days overland; this he took to be evidence of the Far Eastern merchants on the Indian Ocean. In all these conjectures, Columbus maintained the supposition that he had done no more than reach the eastern coast of the known world, even though he was never able to find the passage to Europe explored by Polo.

That inability motivated several other explorations, notably the Portuguese expeditions of Amerigo Vespucci. In 1502 Vespucci set sail for the 'firm lands' which lay south of the Spanish Indies, expecting to find the western passage by persisting southward along the Brazilian coast. He proceeded to about 50 degrees south latitude without success, though it became obvious to him that the coast was indeed a *continental* coast which seemed to expand indefinitely southward. In a series of letters Vespucci made it clear that the land masses he had encountered were *not* part of Asia; in other words, they were not part of the known world. This was a massively important realization, for it contradicted a dominant supposition of medieval Christian civilization, namely that man had been put on *one* world, the 'orbis terrarum', a megacontinent divided into Europe, Africa and Asia. As Edmundo O'Gorman outlines in his brilliant analysis, *The Invention of America,* it was heretical to imagine the earth as anything but a single orbis, for this implied the heretical belief in a plurality of worlds, and a plurality of *man*kind.[19] This pagan notion had been rejected by Christian culture. Nonetheless, Vespucci's explorations, in contrast to those of Columbus, indicated that there indeed was a plurality of 'worlds' and that these new worlds were inhabited. Though this realization was unorthodox, it was nonetheless palatable, because in spite of Columbus' optimistic conjectures about the isthmus, no Asian passage had yet been found. In 1502 and shortly thereafter, cartographers began to represent the new discoveries as large islands separate from the Asian land masses. And in 1507, Martin Waldseemüller's historic world maps were published in the *Cosmographiae Introductio* by the academy of St Dié; the maps illustrated the new land masses as a 'fourth' part of the world separate from Asia, Africa and Europe, and were accompanied by a Latin translation of some of Vespucci's correspondence.[20] In tribute to Vespucci's conception of the idea that these were *new* parts of the world, the cartographer named the southern land mass America, a feminine form of Amerigo's name. With the circulation of the *Cosmographiae* and the continuing exploration of the new worlds, it became customary to refer to *all* the new lands as America, not just those initially explored in the southern hemisphere by Vespucci.

In light of this historical development, to claim that 'Columbus discovered America' is really a misleading observation. Certainly Columbus led the initial expeditions westward across the Atlantic; but that it was 'America' which was discovered was something made evident only with the passage of time. With Vespucci's recognition that the Indies were part of an entirely new hemisphere, Columbus' earlier efforts were retrospectively revised. The histories prepared in the mid sixteenth century all reconstruct the discovery of America with Columbus, uniformly overlooking the fact that at the turn of the fifteenth century, Columbus mistakenly believed he was in a part of the world entirely different from his actual location. Consequently, though whatever he was doing was unprecedented, the nature of the achievement itself (that this was, after all, 'America' which had been discovered) had to unfold over a period of fifteen years, as later explorations objectified earlier ones retroactively. So to return to an earlier discussion, discoveries are not just static 'perspectives' on an event, but are retrospectively and prospectively organized and objectified social statuses.

These observations are not limited merely to geographic discoveries like the discovery of America. The same issues emerge in natural science, as can be seen in the debate over the discovery of oxygen.

DID LAVOISIER AND/OR PRIESTLEY DISCOVER OXYGEN?

Passing reference was made earlier to Kuhn's discussion of this case. It was noted then that for Kuhn discovery is based on the researcher's perception of anomalies in the course of scientific investigations. The overall strategy of such an explanation was identified as *mentalistic* in that it focuses principally on gestalt switches in perception. However, regarding the discovery of oxygen, in Kuhn's judgement such mentalistic considerations seem to be adumbrated by a more common sense criterion. We are told that 'at least three men have legitimate claim to it'.[21] However, Kuhn immediately qualifies this statement. 'The earliest of these claimants to prepare a relatively pure sample of the gas was the Swedish apothecary, C. W. Scheele. We may however ignore his work since it was not published until oxygen's discovery had repeatedly been announced elsewhere.'[22] Whatever the theoretical appeal of the concepts of paradigm and anomaly, Kuhn's reference to 'claimants' and 'claims' reflects the common sense notion of discovery as a unique or unprecedented achievement produced by some subject. That is, the discovery of oxygen is conceived of as a matter of priority in which there are various claimants whose records can be read as demonstrations of their experimental identification of 'empyreal air' (Scheele), 'de-phlogisticated air' (Priestley), 'air itself entire without alteration' (Lavoisier), etc. Hence it is taken for granted that the definition of discovery contains an essential element of novelty, and that the element of novelty is decided through the matter of publication or announcement and the relative timing of such announcements. However, this consideration, even when limited to Lavoisier and Priestley, presents Kuhn with problems as witnessed in the following:

Was it Priestley or Lavoisier, if either, who first discovered oxygen? In any case, when was oxygen discovered? In that form the question could be asked even if only one claimant had existed. As a ruling about priority and date, an answer does not at all concern us . . . Discovery is not the sort of process about which the question is appropriately asked.[23]

Kuhn's indifference to the matter of priority cannot be treated seriously, especially after his exclusion of Scheele, Bayen, Hales and others. Indeed, the remainder of Kuhn's exposition consists of deciding answers to these questions.

Two problems confront Kuhn in his presentation of this discovery. The gas that both Priestley and Lavoisier isolated by burning red oxide of mercury, although called 'oxygine' in Lavoisier's *Traité Elémentaire de Chimie* (1789) (and in his memoirs to the French Academy in 1777, read in 1779 and initially published in 1781), was not understood by Lavoisier and Priestley in the manner in which it is understood today. For Priestley, oxygen was dephlogisticated air.[24] For Lavoisier, it was understood as an atomic 'principle of combustion', that was produced only when the 'principle united with caloric, the matter of heat':[25]

he assigned to dephlogisticated air the name oxygen or 'acid producer', on the erroneous supposition that all acids were formed by its union with a simple, non-metallic body. Combustion was explained by Lavoisier not as the result of liberation of hypothetical 'phlogiston', but as the result of the combination of the burning substance with oxygen.[26]

Hence we see that although Lavoisier and Priestley, as well as Scheele and numerous others like Hales, Black and Bayen, produced relatively pure samples of the element which today any chemist would recognize as oxygen, the various identities given to the substance were entirely different. It would seem more accurate to treat Lavoisier, in particular, not as the discoverer of oxygen, but as the discoverer of the principle of combustion, for his understanding of oxygen as the foundation of acidity and his belief in the weightless, invisible matter called 'caloric' is almost as arcane as the theory of phlogiston from which his rival, Priestley, never broke.

The question 'Who discovered oxygen?', like the question 'Who discovered America?', erroneously implies an object constancy existing between the present conception of the phenomenon, and the earlier conceptions. For the modern chemist the identity of oxygen is a different matter with different guiding assumptions (i.e. the theory of atomic weight and the periodical table versus phlogiston and caloric matter). As Kuhn suggested earlier in his reference to Stephen Hales, to make a discovery, one must be apprised of the gravity of one's achievement; yet in this case, Lavoisier, though compelled by the value of his research, never really determined what for us that in fact was. Nonetheless Lavoisier seemed quite taken by the success of his own work. That is, his disclosures occurred very definitely in the course of research on the problem of combustion and resulted in a locally successful theory. Hence the social basis of the discovery was enforceable endogenously in the research setting itself. As in the case of Columbus, Priestley and Lavoisier were aware of the fact that their work was situated in a context in which its completion would constitute a successful, socially recognizable advancement in science. That was certainly the case with Columbus. It also characterized the interaction of Crick and Watson reported in *The Double Helix*.[27] In this sense Kuhn's suggestion that to discover something, one must be aware of the significance of the achievement as a discovery reflects one of the criteria identified earlier: that discovery is an achievement situated in courses of research action. On the other hand the element of self awareness seemed to be absent from the laboratory of Hales, whose work appears to have been directed toward discovering and perfecting *methods* of isolating different discrete gases – including oxygen. Hales seemed to be unconcerned with what the theoretic significance of his substance could be. Similarly Columbus, while motivated in his explorations, was unaware of the real significance of the Indies. In a parallel way, Kuhn appears to gloss the differences between Priestley and Lavoisier and between them and us by imputing the modern sense of O_2 to their work; hence he retrospectively objectifies *their* enterprises as courses of action directed toward what we today call oxygen. He does not do the same for Scheele because, by the time the latter comes to light, he has lost the race *locally*, or so it seems.

The second problem for Kuhn is *the date* of the discovery. The experiments of Scheele, Priestley and Lavoisier took place over a period of about ten years. This temporal indefiniteness is one of the elements which no doubt prompted Kuhn to disclaim any interest in priority. Nonetheless we cannot fail to be impressed by his out of hand dismissal of the relevance of Scheele. Though an apothecary by training, Scheele was nonetheless a highly prolific scientific writer and researcher. Scheele identified not just the novel properties of oxygen, but a whole host of other new elements; his writings on oxygen disclose not one, but ten separate methods of generating and isolating this new element, 'empyreal air'.

Kuhn's dismissal of Scheele appears even more problematic when we consider the conclusions of Uno Bocklund, who reports evidence to the effect that Scheele may have written to Lavoisier requesting of the latter that he perform an experiment on the burning of mercuric oxide in a vacuum with the aid of the large magnifying glass available in the French Academy of Science. If Bocklund is correct, Lavoisier may have been put onto the investigation of oxygen as a result of the insistence of Scheele, who without the Paris magnifying glass was unable to conduct the crucial experiment himself.[28] At any rate, Lavoisier never gave any hint of this alleged correspondence from Scheele, and few historians consequently credit him with any role in the development of the theory of combustion. His name does nonetheless appear in many of the lists of multiple discovery for the discovery of the element oxygen.

To return to the issue of dating, Kuhn's seeming disinterest in the date of the discovery does not derive from a principled conception of discovery as a matter of social recognition: this position would make the date a function of the groups conferring the status and the attributional processes underlying such an event. On the contrary, Kuhn's position appears to derive from his understanding of discovery as a 'complex event' that develops over a period of time. He contrasts this conception of discovery as a complex event with the admittedly misleading idea implied by the assertion 'oxygen was discovered', which suggests that 'discovering something is a simple act assimilable to our usual (and also questionable) concept of seeing. That is why we so readily assume that discovering, like seeing or touching, should be unequivocally attributable to an individual and to a moment in time.'[29] The present examination of discovery certainly would concur with this observation. However, what understanding does Kuhn extend in opposition to this admittedly romantic view? First, he argues concretely that *two* persons are involved, Lavoisier and Priestley (e.g. 'Lavoisier started the work that led him to oxygen after Priestley's experiments of 1774, and possibly as a result of a hint from Priestley'). And second, continuing in the same vein with regard to the date: 'Ignoring Scheele, we can safely say that oxygen had not been discovered before 1774, and we would probably also say that it had been discovered by 1777 or shortly thereafter.'[30] Hence Kuhn addresses the equivocal matter of deciding *who* discovered oxygen, and *when* it was discovered, by judiciously narrowing down the claimants and subsequently limiting the temporal parameters. These procedures, no matter how loose, indicate that the discovery of oxygen for Kuhn was nonetheless a matter of the 'complex' mental appropriation of some natural facts, as opposed to the social construction accorded to the announcement of such events.

In this respect Kuhn's judgement does not completely break with the folk understanding. His treatment of Scheele only reinforces this.

However, Kuhn does exhibit a sympathy for the sociological view in a separate passage when he states that the nature of discovery lies, at least in part, in the processes of *recognition* associated with a scientific finding; he says 'that if oxygen were dephlogisticated air *for us* we would insist without hesitation that Priestley had discovered it'.[31] Although the specific point of the example is to propose, like Selye and Fisher, that discovery is a matter of viewpoint or perspective, the more general implication is the role of the social recognition in the constitution of a phenomenon's identity. As noted earlier such a proposal misses the locally compelling nature of discovery. Nonetheless, it turns our attention away from the mentalistic processes in the perception of anomalies to the structure of recognition by virtue of which some solution to those paradigmatic anomalies is deemed a discovery in the course of research.

It is apparent that discoveries are social events whose statuses as discoveries are retrospectively and prospectively objectified; however, it is also clear that members of society orient to these statuses as though they were ordinary social facts which take their identities from their 'natural' or inherent value. In other words, members of society in one instance socially construct an event as a discovery only to later orient to it as a natural fact of life. One final example will reinforce the interpretations made in our examination of America and oxygen.

The case of the Piltdown Man is particularly compelling for our purposes, because it has experienced a very turbulent scientific career from the time of its initial announcement to the contemporary debate over its actual status. In each case its status is clearly a production of the methods of recognition brought to bear on it. In this respect it strengthens the force of the other cases studied here.

THE DISCOVERY AND THE CAREER OF PILTDOWN MAN

In 1859 Darwin announced in *The Origin Of Species* that plant and animal species evolved in successive forms through geological time to the forms in which they appear in present times. Twelve years afterwards in the *Descent of Man,* he furthered this line of thinking in his speculations on the common evolutionary roots of man and the higher apes. At the time he was writing, only fossil remains of the Neanderthal Man had been uncovered (1857). Yet Neanderthal was a very advanced evolutionary specimen, greatly resembling modern man.

Darwin postulated that some kind of missing link would be discovered which would bear out the common evolutionary origins of species of man and species of apes. The first such important find in Europe was the Mauer Jaw of Heidelberg Man. This was the first early progenitor of modern man, uncovered in 1907. The Mauer Jaw is ape-like in the chin area, but the teeth which were found preserved (though dislodged) were very human in form. Several years later at the meetings of the Geological Society of the Royal Museum in London a comparable specimen was brought to light: 'On 18 December 1912 Arthur Smith Woodward and Charles Dawson announced to a great and expectant scientific audience the epoch-making discovery of a remote ancestral form of man – the Dawn Man of Piltdown'.[32] The remains were proclaimed to be the predecessor of the Mauer Jaw. They included a large section of skull cap and one side of the lower jaw, as well as some remains of extinct animals – mastadons, stegodons, rhino – and evidence of very primitive tool-making. The stratigraphical evidence suggested that the remains were from the lower pleistocene era, or the late pliocene – perhaps some 500,000 years old.

The striking feature of the find was that the skull seemed very large and hence concordant with a well developed brain, while the jaw was very thick and ape-like, though the two molars were flat and moderately human.

Here indeed was a candidate for the missing link, with clear evidence of both human and simian features. Dawson had written to Woodward in February 1912 that the find was 'part of a human skull which will rival Homo Heidelbergensis'. Similarly Arthur Keith, a leading anatomist, observed that 'the discovery had fulfilled the prophecy of what the ancestor of man was likely to be'.[33] The size of the skull in association with the other remains made the announcement of Piltdown Man altogether more significant than the earlier discovery of the Mauer Jaw. It was named, in honour of the discoverer, *Eoanthropus dawsoni*. Dawson's 'dawn man' was a specimen which emerged against a background of expectation initiated by Darwin and corroborated by the Mauer Jaw.

in eoanthropus dawsoni, we seem to have realized a creature which had already attained to human intelligence but had not yet wholly lost its ancestral jaw and fighting teeth . . . [it was] a combination which had indeed long been previously anticipated as an almost necessary stage in the course of human development.[34]

However, immediately upon presentation of the results, there were objections. Some, like David Waterston and Arthur Keith, were much impressed with the size of the skull – which they took to be evidence of its probable recent origin. Yet they were perplexed by the primitiveness of the jaw, and its apparent co-presence with the pliocene fossil animal remains. They reasoned that the artifacts could only be made sensible by assuming that the more ancient remains had been washed into the Piltdown quarry site from an earlier deposit, and that the jaw belonged to a separate creature, an extinct ape. Woodward retorted by showing that the molars exhibited were peculiar to those found in man, and he predicted that if and when a canine or eyetooth was found, this would show human development and would probably not obtrude above the level of the other teeth, as it does in simians.

Clearly what one assumed to be found, or rather what one recognized, determined how one ordered the particular elements composing the find. Waterston and Keith recognized 'in' the remains that the skull was recent: hence the remains were read as an exhibit of displacement in which the contiguity of the elements was brought about accidentally. Yet for Woodward and Dawson the remains were undeniably primitive; the erosion of the molars was distinctive evidence of hominoid mastication, and consequently their association with the advanced skull was incontrovertible. Just as Waterston and Keith glossed the association as accidental and reconstructed the geological circumstances which could have brought about just such a display, so too Woodward reconstructed the palaeontological circumstances to provide for the coherence of the remains, and to allow for the probable structure of future remains. In either case, whatever composed the find was glossed to reconstruct the events so as to show how natural forces could have operated to produce what the archaeologists were confronted with.[35]

Not surprisingly the identity of *Eoanthropus dawsoni* was confirmed for Woodward, Dawson and associates by later digging. In 1913, Teilhard de Chardin, working with Dawson, found the all-important eyetooth 'close to the spot where the lower jaw itself had been disinterred'.[36] Had it been made to order, it could not have been closer to Woodward's expectations. Also, a year later digging uncovered what appeared to be a carved slab of elephant tusk; and in 1915 in a field site at Sheffield Park two miles from Piltdown, Dawson uncovered the cranial remains of another Piltdown skull, together with a molar tooth, and a tooth from an early pleistocene rhinoceros. Whatever doubts were expressed regarding the age of Piltdown Man, the functional unity of the jaw and skull, and his prowess as a toolmaker were dismissed by the new findings. The crudity of the jaw and its association with the ancient animal fossils recommended that here was a very early ape man, in fact the earliest to date. This opinion reigned, with minor exceptions, for some twenty or more years. Dawson, the country lawyer and amateur palaeontologist, received widespread acclaim, and twenty years after his achievement Sir Arthur Smith Woodward, the keeper of the Department of Geology at the Royal Museum, who had brought Dawson's work to the attention of the scientific world, erected a memorial stone for Dawson at the gravel site which was unveiled in 1938. Dawson had died decades before in 1916, shortly after the last discoveries had been made.

In about 1936 a whole series of new discoveries of pithecanthropine man were made in Africa, China, and Java. The earlier claims of Dubois in 1891 regarding the human form of Java Man were vindicated by the discovery of a new specimen which was almost a duplicate of the first. A more advanced, but closely allied species, the fire-making, tool-using, cave-dwelling Peking Man, was also discovered and located in the early ancestry of man. Furthermore, Dart and Broom were coming upon perhaps the most primitive and earliest forms of hominids, the australopithecines, in South Africa. Whereas the Piltdown specimens portrayed the earliest man with very *simian* jaws and *human* skulls, later research indicated just the opposite – very human jaws associated with more primitive skulls. These discoveries suggested two very separate lines of evolutionary development which included *Piltdown Man in the one line of descent, and all the other early men in the other*. Needless to say, the archaeological community was very unsettled with this situation.

In 1949 Kenneth Oakley sought to put these matters straight by dating the Piltdown remains with a newly discovered dating process. The results were staggering. Both jaw and cranium were found to have such low levels of fluorine build-up that they were estimated to be no more than 50,000 years old. This meant that the Piltdown Man was a late ice age curiosity; and since he postdated some already existing forms, he was of no consequence in their development.[37] It was also problematic to attribute such crude tools (palaeoliths, flints) to so apparently recent a hominid. And if the jaw was attributed to an ancient ape, as some still believed, it was difficult to reconcile the fauna associated with the ice age beaver remains with that of ice age apes, especially since the fossil apes were mostly restricted to Africa and Asia. Following such disclosures, Oakley was persuaded to conduct an intense examination of all the artifacts associated with Piltdown.

In 1953 as the result of a barrage of chemical, physical, and anatomical tests on all the artifacts, the Piltdown forgery was disclosed.[38] Oakley, working with J. J. Weiner, determined that the jaw had been taken from a modern female orangutan, and stained brown with chromium; the canine likewise had been obtained from a recent orangutan, but had been filed flat and painted with Vandyke Brown paint; the molars were ape molars filed flat and dyed with chromium; although the skull proved to be late ice age, it too had been dyed like the other artifacts; the carved elephant tusk was palaeolithic, as were the mastadon, stegodon, and rhino teeth, but all had been chromium dyed to appear contemporaneous with the skull and jaw bone. Artifacts had been 'salted' in the Piltdown site to later appear as *in situ* findings to be recovered in the digs. Hence the Piltdown Man was said in fact to be a forgery. That is, the recognition of fraud was now used to order the elements of the case.

Certain facts indicated that Dawson, the amateur palaeontologist, who had first uncovered the fossils, had engineered the whole affair. Choosing the specimens from the excellent collection he had gathered and bought, he dyed and filed them appropriately, presented them to Woodward at the British Museum claiming to have discovered the missing link, and consequently, as the need arose, planted further appropriate specimens in the sites to suit the growing expectations of the community. When he died in 1916, four years after the announcement of the missing link, the last 'discoveries' had been made. In 1950, a large new section of the Piltdown quarry was opened up, and tons of earth and gravel were sifted and examined; not a single fossil or artifact was uncovered. Dawson's handiwork in the earlier successful excavations had been fairly well established.

The story of the Piltdown forgery is instructive in our examination of the social dimensions of discovery. The Piltdown remains became candidates for a discovery because the understanding of antiquity expounded by Darwin and corroborated by the Mauer Jaw led the scientific community to expect that some such finding would inevitably come to light. These expectations were instructive to Dawson in providing him with some guidelines along which to constitute his forgery. That forgery only became suspect when the expectations of the scientific community shifted following the later pithecanthropine findings. It was only then that Piltdown became problematic. Hence the 'discovery' was intimately tied to the expectations of the scientific community which were used to gloss the array of elements which composed the Piltdown remains. Psychological models of discovery would assume that the discovery event occurs in the course of an inductive stage, is mentalistically realized, and coterminously verified in the confines of subjective experience. However, in the case of Piltdown Man, the 'steps' which preceded what for forty years was the announcement of the discovery might have included reflecting on the idea of the missing link, inventing a plausible construction of it, preparing the fossil remains, and planting them in a plausible location so as to communicate a version of ancestral man and his tools which was concordant with the very great antiquity attributed to hominid remains.

The planning also included initiating the participation of the scientific community – the apparent object and beneficiary of the whole plan. This step was decisive. Without it, even if the remains had been dyed and prepared as fossil collections to be stored in Dawson's cupboards at home, along with his Barcombe Mills Man,[39] they would have been nothing more than a curio, like the latter. In the context of the 1912 scientific recognition, the artifacts became pieces of a discovery. That is, the notion of discovery was used to order the remains, employing notions of continuity with Darwin's expectations, the present fossil record, and imagined lines of succession with apes on the horizon and European man on the leading edge.

Intuitively it does not seem improbable that the sole motivation for these actions was the public recognition and acclaim which would result for their author. In this respect Dawson's endeavour was fully consistent with Merton's conception of discovery as a ceremonial event in which the scientist is transformed into something of a celebrity or hero.

This is part of the interpretation which emerged with the 1953 *reconstruction* of the original announcement. By treating the event as a forgery, one is providing a scheme of interpretation in terms of which the actions of Dawson are made sensible as actions of a man striving for a legitimate end – social approval – through an illegitimate means – deception. In this account, all the elements of the Piltdown find are 'artifacts' of one person's inordinate preoccupation with his social status – not artifacts of an archaeological scene. This interpretation also explains how it was that the scientific community found itself in this predicament: it took for granted that the materials uncovered were materials derived from an ostensively bona fide course of archaeological research when 'in fact' they were not. Also, the idea that Dawson had perpetrated the fraud is corroborated by the fact that after his death no further finds were made; this would be expected if he were indeed the inventor of the uncovered materials. Significantly, the date of Dawson's death plays no role in the earlier account of the events as a discovery. However, this has not been the last word.[40]

Ronald Millar has re-opened the case with compelling arguments which show that the Piltdown Man was neither a *discovery*, nor a *fraud* – but a *hoax*. Millar assembles profiles of the personal characters of the major parties involved – Woodward, Dawson and Chardin – arguing that none of their personalities or stations in life could be taken as consonant with the perpetration of such a forgery. The guilt lies elsewhere.

According to Millar, certain inconsistencies in the publications of one Grafton Elliot Smith, a comparative anatomist, showed that he must certainly have examined the original fragments when in fact he claimed to have seen only the plaster cast. Also, even though while in Cairo he had conducted an archaeological survey of some 20,000 prehistoric burials at Nubia, and had gained a rare expertise in anatomical reconstructions, he failed to correct the faulty reconstruction of the Piltdown remains by Woodward and his assistant, W. P. Pycraft, though it must have been painfully obvious, even from a plaster cast, that they were mistaken. Neither Woodward nor Pycraft were skilled in human anatomy; Woodward was a palaeoichthyologist, and Pycraft an ornithologist.

Also, according to Millar, Smith had the reputation of a prankster and dearly loved to have a laugh at the expense of his supercilious colleagues. This is offered as the real motive for his actions. But it is also pointed out that perhaps only he had access to all the artifacts involved. The finds uncovered at Piltdown were so rare and unusual that only someone with Smith's experience would have had access to them. Furthermore, the eyetooth which Chardin picked up was later found to be somewhat radioactive, just as were the remains from a site in Tunisia, which Smith visited during his numerous Mediterranean travels. Similarly, with his catalogue of thousands of ancient and prehistoric skulls from Nubia, no one could have had a better selection for such a hoax. Lastly,

Grafton Elliot Smith also told poor Dawson that Piltdown Man's brother or cousin from Talgai was found at a place called Pilton in Queensland. In fact there is no such place in the whole of Australia. Could Smith's eyes have watered just a little as he watched the innocent dupe Dawson swallow this gobbet of false information?[41]

Having established to his satisfaction that the ruse was perpetrated by Smith, Millar directs the reader's attention in the closing paragraph of his book to the famous portrait by John Cooke of the Piltdown personalities. All the principals are present except for Teilhard de Chardin. However, the portrait shows that Grafton Elliot Smith, who otherwise was only of secondary importance to the investigation, is represented centrally in the portrait. Indeed he is pointing out features of the Piltdown skull to Arthur Keith, while all other principals look on attentively. In light of Millar's suggestions, Smith's directive becomes compellingly prescient. Here too is fascinating though ironic corroboration of Smith's handiwork. Millar's interpretation transforms the whole sensibility of the picture.

In this account the reputations of Woodward, Dawson and Chardin are all left unscathed. Each is believed to have been working in good faith, and to have been taken in, not by a malicious fraud, but by a monumental prank. The ruse was never dispelled by Smith because, it is conjectured, so seriously was Piltdown taken by the profession, that his disclosure would have caused such adverse opinion as to have caused him to have lost all scientific credibility.

Each interpretation of the events of Piltdown illuminates its own package of relevant details, details whose integrity or unity is occasioned by the method used to examine the case. In contrast to his earlier failure to pay attention to the artificially cut abrasions which appeared on the Piltdown molars, Le Gros Clark noted that when the suggestion of the fraud was made, 'the evidence of artificial abrasion immediately sprang to the eye'. He continued, 'Indeed so obvious did they seem it may well be asked – how was it that they had escaped notice before? (He answered his question with a beautiful simplicity.) They had never been looked for . . . nobody had previously examined the Piltdown jaw with the idea of a possible forgery in mind, a deliberate fabrication.'[42]

Le Gros Clark's remark encapsulates the whole point of our review: each interpretation of the event attributed to it something of an implicit understanding of how it was brought about, the integrity of the researchers, the likelihood of their success, the continuity of the work, its probable importance, the endemic ambiguities of such a project, the disclosures which could be expected to follow from the present one, etc. In other words, the assumption that the event was a discovery constituted a virtual method of interpretation which was used to explicate the relevant factual circumstances assumed to be co-present with the discovery; similarly for the hoax and the forgery.

In the recognition of the discovery the antiquity of the remains is the forceful overarching fact. This is corroborated by the primitiveness of the mandible, the antiquity of the animal forms, the uniformity of the colouring, the wear of the teeth, etc. In the recognition of the fraud, the personal motivation of Woodward is brought to bear. His earlier finds are regarded with suspicion, and the details of his life are searched for evidence of opportunity, inclination, motivation, etc. In the recognition of the hoax, the mischievous character of Grafton Elliot Smith becomes primary. The good reputation of the principals in the discovery is salvaged. Each account occasions its own corpus of corroborating elements.

However, no account has proven to be definitive. Indeed, recent disclosures have re-opened the entire matter. Before he died in 1978, Professor Douglas, who held the Chair in Geology at Oxford, made a tape-recording in which he alleged that the Piltdown fraud had been concocted by his former teacher and predecessor in Geology at Oxford, Professor W. J. Sollas.[43] According to Douglas, Sollas had been slighted for his synthetic reconstruction of a primitive skull early in the century by Smith–Woodward at a professional meeting. The skull had been removed from its stone casing by grinding away the materials around the bone, making models of the original bone, then grinding both stone and fossil away until more bone was uncovered. Smith–Woodward had dismissed the reconstruction and the months of labour which went into its preparation as a 'toy'. Douglas argues that Sollas was so enraged by this incivility that he planned the Piltdown hoax to discredit Woodward, for though Woodward was held to be a prominent palaeontologist, his talent, in the estimation of Sollas, was merely that of an artist. He could describe specimens wonderfully, but he had no conception of their theoretic relevance. According to Douglas, Sollas hoped to dupe Woodward with his bogus specimens, making him the laughing stock of the scientific community, for it was assumed that the ruse would immediately be recognized by the cognoscenti, especially since the carved elephant tusk was modelled after a cricket bat. The implication was that the earliest Briton played cricket! However, the ruse backfired, the discoveries gained significant community momentum, and Woodward was eventually knighted for his efforts. Consequently, the secret of the hoax was never disclosed by Sollas, and came to light only after his death and the death of his pupil, Professor Douglas. Though it may take historians some time to weigh the claims left in the tape made by Douglas, for our purposes the lesson remains the same. This new interpretation, like the previous claims, objectifies a shifting corpus of elements, transforming the relevance of the past details in light of the present knowledge at hand. Just as the original Piltdown was seen as a fulfilment of Darwin's prediction (which raised its own problems), so too the recognition of fraud solved other problems brought about by belief in the validity of the Piltdown remains. Similarly the accounts of a hoax revise yet other key details. Each successive interpretation brings a sense of coherence to the details, though never with complete success. Even the latest accounts are notably weak on first hand reports of Smith and Sollas' activities. These rely for the most part on circumstantial evidence. What is interesting from our perspective is that each account is given within what Pollner calls mundane reasoning. In contrast to the panorama which the analyst observes, each author holds steadfastly to the belief that he is describing the same unique circumstances. From our perspective, what changes are the methods used to realize the details of such events and the objectified sense of occurrence which each method lends to its interpretation. This explains why each version of the event strikes its advocates as so obvious and natural. However, seen in perspective, we realize that these various 'objectivities' are social accomplishments – however natural and obvious they appear at first sight.

This paradox is what the present chapter has been directed towards. While in academic circles, we often are ready to accept the relativity of customs and beliefs – even our own – we seldom are aware that for members of society involved in their pragmatic affairs, the structures of everyday life are real, obdurate, natural and unavoidable social facts. Discoveries are among such facts. The reflective examination of these discoveries indicates that they are not simply objective, but they are *objectified* facts. In an uncanny and involuntary way they are constantly undergoing a retrospective and prospective interpretation – even here when as analysts we try to study them – as Kuhn's experience illustrates.

However, the objectification of actual discoveries does not end with the reflexivity of their adoption. Not only do theories *of* science reify their factual domains, but theories *about* such theories act in a similar manner. This is the subject of our next chapter.

8

Folk reasoning in theories about scientific discovery

Numerous students of science have drawn attention to certain 'ritual' features of scientific behaviour. For example, Paul K. Feyerabend talks about science as the scientists' religion, and suggests that scientific change often occurs in the same way as religious change: through rhetoric, charisma, and appeal to the temperamentally sympathetic. Also, Hagstrom has developed the parallel between publication in the scientific community, and rituals of 'gift giving' in pre-literate societies. More recently certain historians of science have begun to uncover a different theme: the mythologizing or 'apotheosis' of discipline founders. Paul Forman suggests in his analysis of the history of X-ray crystallography that, in the interests of maintaining and strengthening a separate disciplinary identity, physicists in this area have inadvertently reconstructed the origins of their field with little regard for the historical facts and have produced in their retrospective histories 'an account of the conceptual situation in physics circa 1911 which is, in certain respects, utterly mythical'.[1] Robert Olby argues that a similar 'mythicization' occurred during the 'Mendel revival' at the turn of the century when the proponents of the new field of 'genetics' (a term coined by Bateson) 'aggrandized' the achievements of Mendel which had appeared some thirty-six years earlier. Hence, in his controversy with the biometricians, Bateson could not only point to the ostensive superiority of his model, but could point to what he held to be the critical collaborating discoveries of Mendel which had been 'overlooked' in the historical record until 1900. Olby then explores the evidence supporting his conviction that 'Mendel was no Mendelian'[2] – a conclusion which reinforces the analysis offered here.

The present analysis examines not myths of origins of specific scientific fields, but certain beliefs about the processes by which discoveries are made. Specifically, it is suggested that two prominent scientific accounts of scientific discovery which we encountered frequently in the first few chapters are in fact folk beliefs whose value derives from a prominent phenomenological assumption regarding the 'natural world'. It is suggested specifically that in spite of the fact that they

143

are easily dispelled on a number of grounds, these two theories have enjoyed enormous currency because they resolve certain dilemmas posed by the juxtaposition of the phenomenon of discovery on the one hand with the reciprocity of perspectives on the other.

Folk elements in theories of science and theories about science

Historians of science are familiar with numerous cases of 'discoveries' which were possible and which were recognized only because of the expectations of the communities in which they were made. We have just examined a more notable case: the Piltdown forgery. The initial resistance to this 'Dawn Man' of Piltdown by palaeontologists like David Waterston and Arthur Keith was ignored as the great majority of the scientific community in Britain endorsed the discovery in what now appears to be an almost nationalistic fervour. While the Germans had their Mauer Jaw of Heidelberg Man, the British now had their own 'missing link' with its human cranium and simian mandible. Though this forgery was most likely not concocted out of Victorian chauvinism, it clearly enjoyed a symbolic status in the community once it came to general attention.[3]

Similarly, recent evidence indicates that Cyril Burt's investigation of the genetic basis of I.Q. in his famous twin studies was fabricated.[4] Commentators have argued that Burt probably 'invented' the sample out of his own common sense opinion that the sons of ditch diggers were less well endowed intellectually than the sons of professors, lawyers and doctors. This common sense theory went unchallenged and unchecked for so long precisely because it had a strong intuitive appeal to the members of that class who had come to their positions in the scientific community, presumably via their native abilities. Only in retrospect can we see the significance of Burt's ties to the eugenics movement of early-twentieth-century Britain.[5]

Another case in which 'unconscious finagling' distorted the conclusions of a scientific study has been reported by Stephen J. Gould.[6] Gould re-analysed the data published in Samuel George Morton's 1830s study of the relative cranial capacity of different groups of the human race. To no one's surprise, Morton concluded that the largest average cranial capacity (and hence I.Q.) was found among Caucasians, followed by Indians, and then black peoples. Re-analysis shows that there is actually no mean difference between any of these groups, and that Morton arrived at his conclusions, not via the conscious manipulation of data, but through a series of unconscious slips, inconsistencies, omissions and miscalculations. This leads Gould to speculate that the unconscious manipulation of data may be a scientific norm, and not an entirely exceptional phenomenon.

These cases suggest that scientific conclusions often have a symbolic value above their actual scientific content, and that this symbolic significance often determines the value and even the content and conclusions of research. In other words, from retrospect, historical theories often have obvious folk dimensions.

While it has been argued that all theories of science have a conventional character, I would like to suggest that certain theories *about* theories or discoveries are not only conventional, but have specific cultural consequences for the objectification of the things which they purport to explain, even though they are mistaken. In other words, theories or observations which constitute discoveries objectify a field of experience, but theories *about* discoveries sediment the experience at a second order of abstraction. And while both theories and theories about theories are conventional, the latter are of interest, not because they are conventional, but because they are wrong. Specifically, this book has examined mentalistic and cultural models of discoveries. While I have endeavoured in earlier chapters to describe reasons for rejecting these models, in this chapter I would like to suggest some reasons why they are valuable in everyday life, though not as *theories* for us, but as *beliefs* for members of society. However, since they seem to operate as virtual theories, I shall refer to them as folk or ethnotheories.

A preliminary articulation of the elements of a folk or ethnotheory might include the following. First, it has the status of a rational explanation and is typically the outcome of a course of research action conducted by persons in their capacity as researchers or scientists. Second, the explanation has wide intuitive appeal and is frequently invoked though often in an ad hoc manner. Third, the theory at least from retrospect is objectively incorrect, yet was upheld because it fulfilled some social function for which it was particularly adequate. Cyril Burt's I.Q. studies and Morton's crania studies are clearly folk theories. However, these sorts of examples are adequately described under the term 'ideology'; such theories legitimate the social positions of the individuals, classes or countries who develop and consume them. But folk theories include other types of beliefs in addition to what political analysts have termed ideologies. Some folk theories cut across class, ethnic and national borders and function at the level of common sense. Two such theories are the explanation of discovery by arguments of 'cultural maturation' and by arguments of 'genius'. This chapter will describe how these theories, though scientifically erroneous, have enjoyed frequent and compelling citations because each in its way salvages the incorrigible assumption of the reciprocity of perspectives when this appears to be challenged by acts of discovery.

MULTIPLE DISCOVERY AND CULTURAL MATURATION: THE FIRST THEORY

As noted in chapter 4, Alfred L. Kroeber in 1917 published his famous treatise, 'The superorganic', in which he challenged the attempts to reduce social evolution to a simple biological model. Kroeber argued that changes occur in societies as the result of *cultural* evolution, not species evolution. His position was aimed specifically at Francis Galton who had attempted to explain all of the achievements of higher civilizations by the 'hereditary genius' of the groups in question. According to Galton, great composers, great scientists, and even great wrestling talents appear to run in families over several generations. Ergo, achievement is controlled genetically, and progress in societies is a function of hereditary achievements. Kroeber argued contrarily that in science great discoveries are not acts of individual achievement, but are a function of a 'critical level of development'. His evidence was a list of some fifteen or sixteen multiple, simultaneous, independent discoveries in science.[7] When a culture reaches a critical point, a discovery occurs not just once, but several times over. Kroeber pointed to the rediscovery of the Mendelian ratios and remarked that the triple rediscovery was proof of the ripeness of its time, and that the long neglect of Mendel was evidence that he had anticipated his time.[8]

The original list of multiples was expanded in 1922 by Ogburn and Thomas who published a description of 148 multiple discoveries. This list has become a cornerstone in the sociology of science, leading certain people to argue that *all* discoveries are 'in principle' multiple discoveries.[9] The list has also figured importantly in Leslie White's efforts to establish the concept of 'culturology'.[10] For anthropologists like Kroeber and White, the list seems to demonstrate that changes occur in societies not as the result of genius, individual initiative, conscious planning, and free will, but through an inexorable process of historical maturation.

Such a conception of change directed attention away from the individual researcher to the social context. Hence the possibility of reducing anthropology to a biological model, in which genetically controlled 'ingenuity' determined individual and social achievement, was rebuffed. Indeed, Leslie White was later to refer to ingenuity as a rather common, and hence uncritical ingredient in the precipitation of change in science. Recall that he suggested that Urey's isolation of the heavy isotope of hydrogen required about as much ingenuity as opening a recalcitrant jar of pickles.[11] His major datum for this refutation of genius theory was the list of multiples identified by Kroeber, and expanded and classified by Dorothy S. Thomas. Since the argument of cultural maturation repeatedly relies on the phenomenon of multiple discovery, a closer inspection of the 1922 list is warranted. In what follows, I will entertain an alternative hypothesis to explain the phenomenon of duplication, and will suggest a series of problems which are typically overlooked in the discussion of multiples.

Multiple discoveries and the failure of communication

The mathematician, Jacques Hadamard,[12] noted that mathematicians frequently monitor the work which their colleagues do so as to avoid duplication. 'After having started a certain set of questions and seeing that several authors had begun to follow the same line, I happened to drop it and to investigate something else.' Had Hadamard not seen that others were conducting similar research it is not improbable that both he and they would have continued in their research and subsequently reported rival outcomes. The first datum which Merton[13] cited as evidence for the contention that all discoveries are *in principle* multiple discoveries was that published singletons often turn out to be multiples. This was illustrated with the examples of Cavendish and Gauss. Their original discoveries had been recorded in unpublished notebooks. Is it any wonder that others subsequently failed to take note of such achievements? Presumably, if such works had been published or communicated widely, the potential rivals would have worked in other directions and consequently the record of multiples would have been truncated significantly.

This was intimated by Ogburn[14] in his work on *Social Change*. He stated: 'if an invention has become once made and has become widely known, there is no occasion for a second invention'. Of course, the reason for this is that once an invention or discovery has become known, the appearance of any identical announcement will be treated as a replication or duplication whose identity will be recorded in comparison with the earlier unprecedented discovery or invention. However, the announcements of discoveries do not always meet with unqualified success. Merton[15] spoke of the possibility that an announcement could get 'lost in the great information system' of science. Presumably a discovery could get lost or could fail to become widely known or could be announced well after the achievement had been made elsewhere, because it was not published or circulated (this was the case with Scheele, Cavendish and Gauss); because it was a matter of political or industrial secrecy (the Manhattan Project); because it was politically or religiously repressed (Galileo in Rome, and Vavilov in Russia); because it was unavailable to others outside the language and/or cultural group in which it was recorded (this was the case until recently with ancient Phoenician stone writing found in North America); or because it was announced in such obscure language that potential readers were disentitled (as in the case of Copernicus, Semmelweis and Galois). Numerous examples could be presented to substantiate each point. Nonetheless the common thread uniting each case would be what might be called a failure in communication. We might argue therefore that the history of science is not essentially a history of multiple discovery, but a history of poor communication. Consonant with this view is the observation that priority disputes, which are intrinsically associated with multiples, have *declined* in time with the rise of professional scientific establishments,[16] establishments which are particularly efficient in organizing the communication of research outcomes.

It does not seem inappropriate to conclude that if communications had been maximally efficient, the incidence of duplication would have been much smaller than it actually is. For the sake of argument we might say that the history of science would be a record of singletons.

Nonetheless, the culturologists would still want to conclude that the time had come even for such single discoveries; their appearance would be proof of the ripeness of their time. However, the claim that the history of singletons originates inevitably with the readiness of culture would be plainly fatuous. Ergo, the apparent efficacy of such a list is available only by ignoring the replications which were tied to failures in communication. Hence the argument of cultural determination, based on such lists, though it may be rhetorically compelling, is nonetheless circular.

Even if one could develop a possible list of single historical discoveries, what could be made of the claim that their occurrence was 'inevitable'? The force of the argument from multiples is that discoveries occur in an *inevitable* fashion with the maturation of culture. This explains why so many duplications have emerged: the cultures had reached an adequate level of preparation. The preparation of a community to produce a new work seems to include notions about the technical efficiency and the theoretical maturity of a science. For example, advanced astronomical observations are made possible by such things as the Renaissance, the harnessing of electricity, the development of metal alloys for the construction of parabolic disks, surplus capital and government support of research, the invention of numerous electronic devices, especially radar and radio receivers for decoding noise from space, the theoretical development of astrophysics, high energy particle physics, and all the rest. Clearly, these developments make future discoveries *possible* by providing materially for their achievement, and theoretically for their sensibility. This however, does not make them *inevitable*. Edward Sapir[17] suggested a better candidate for this.

Dr Kroeber chooses his examples from the realm of inventions and scientific theories. Here it is relatively easy to justify a sweeping social determinism in view of a certain general inevitability in the course of the acquirement of knowledge. This inevitability does not altogether reside, as Dr Kroeber seems to imply, in a social 'force' but, to a very large extent, in the fixity, conceptually speaking, of the objective world. This fixity forms the sharpest of predetermined grooves for the unfolding of man's knowledge.

This can be interpreted to mean that inasmuch as theories provide for certain interpretations of nature, and inasmuch as these can be entertained via concrete observations, it is inevitable that men working with such theories and making such observations will find what they are searching for, and/or will consequently revise their theories and, 'conceptually speaking', re-formulate or re-fix the structure of nature. This understanding is quite consonant with Kuhn's idea of paradigm. However, Kroeber and his followers seem not to have meant anything as analytical as this.

Consequently it would be wiser to conclude that cultural development makes certain classes of discoveries *possible,* and given the existence of accurate research paradigms or programmes, certain discoveries could be *highly probable,* but *never inevitable,* especially on the simple premise of cultural preparation.

There is a further weakness to the idea that discoveries are inevitable; if this idea were true, it would be impossible to account for the existence of controversy in historical science. In his discussion of Darwin, Kroeber[18] suggested that the 'immediate acceptance' of speciation was proof of the 'readiness of the world' as opposed to the 'intrinsic truth' of the concept. From his point of view, the preparation of the culture is evidenced by the favourable climate of opinion in which the theory is announced. However, in fact Darwin's announcement met with such a howl of resistance that one would have expected Kroeber to have claimed that Darwin was indeed *ahead* of his time. In other words, the argument from cultural preparation which suggests that discoveries are inevitable erroneously assumes that discoveries are typically announced where the social context is favourably disposed. Aside from holding the unlikely position that controversies occur when *their* time has come, the argument is far from substantiated by the historical record.

Sources of ambiguity for the record of multiples

I will here speculate on how seriously the 1922 list of 148 multiple discoveries would be reduced if one could actually determine how many cases were the results of duplication that under conditions of better communication might not have occurred. Then I will examine how strong the case for multiple discoveries is when based on the details of that portion of the list not affected by the first assumption.

The limiting case for true simultaneity would be the situation in which two independent researchers announced the identical discovery in the same breath. This has never occurred. Consequently, there has always been some interval between even the most representative cases of simultaneity. The possible period of overlap between two identical researches must be a continuum between *absolute coincidence* and *complete disjuncture,* with the culturologists stressing the former. However, an estimate of the pattern of simultaneity can be made by assuming that discoveries announced in *separate years* tend to be more from that proportion of the record which approach disjuncture – and whose duplication would most readily have been forestalled by speedy and general publication. On the contrary, those announcements which occur in the *same year* would have a greater tendency to represent cases of pure and unavoidable coincidence.

If we delete from the 1922 list those entries which were discovered over a period of more than a year, that is, which were recorded in at least two different years, and whose wider acknowledgement might have prevented duplication, what proportion of the list is affected? The results show that the history of science does *not* become a history of singletons, as we suggested for the sake of argument above, but we can remove well over one half of the entries.[19] Indeed, the majority of announcements are separated by more than one year. It would be difficult to argue that the history of science is *in principle* a history of *multiples* when there appear to be less than seventy-five instances. Indeed it would be most difficult, in light of the fact that this group of entries is plagued with major difficulties. Specifically, it includes as independent multiple discoveries cases in which one party accused the other of having stolen his ideas. It includes cases in which Thomas herself expressed doubts about the legitimacy of certain of the claimants. It includes cases in which the dating was ambiguous and cases in which the contributions though concurrent were radically different in quality and completion. It includes as well cases of discovery whose significance was not locally recognized, but was retrospectively reconstructed.

For example, the discovery of sunspots is listed as the seventh item by Thomas. The researchers that she identifies are Galileo, Fabricus, Scheiner and Harriott. All are credited with the discovery in the same year, 1611. Although we now have independent evidence that Fabricus had published such observations first, Galileo in *The Assayer* accused Scheiner of having stolen the ideas from him, and of pretending to have arrived at them independently. The problem of plagiarism is that in most cases we have no definitive way of knowing whether it has actually occurred, and if we grant that charges of plagiarism recorded in the historical record are well founded, it may well be that these suspicions were voiced only when the 'borrowing' has been so explicit that it was evident to the offended party. Indeed, its real incidence may be far greater. Thomas also points out that credit for the invention of certain famous instruments and tools, for example, the microscope, telescope, thermometer, steamboat, electric train, and telephone, are *still* matters of dispute, as she acknowledges with the qualifier 'claimed by' after several of the entries.[20] This ambiguity is aggravated by the fact that the dates for many of the simultaneous achievements are uncertain. In other words, in a number of the entries, many of the personalities as well as the dates are the subject of doubt.

Yet even when the dates and personalities are certain, there are often enormous differences in the relative quality of the contributions of each party. For example, Thomas points out that though Halley and Newton both worked on the law of inverse squares, Newton's efforts were far superior, though both are given equal credit by the list. As I have indicated in chapter 6, this problem is even more complicated in the research associated with the discovery of genetics. Though De Vries is credited as a rediscoverer of Mendel's Laws in 1900, it has since become evident that he thought Mendel's account was an *exception* to the general rule,[21] and that genetic variability was produced principally by mutations. Also, Tschermak, who was the third of the Mendel rediscoverers, was *not* working on the problem of heredity in his research, but on the agricultural problem of xenia in hybrids.[22] Xenia is the effect of cross-fertilization on the fruit or endosperm of hybrids. For him, hybridization was of importance in developing new endosperm products, not as a clue to the mechanism of inheritance. In fact, he only realized the relevance of his own studies *after* reading the reports of Correns and De Vries. For this reason, authorities like Curt Stern have not included him on their lists of the rediscoverers of Mendel.

One finds this issue in other examples. Kuhn[23] observes that while there were many researchers working with oxygen, all can be ignored except for Lavoisier and Priestley. He discusses only these two, while Thomas lists Scheele and Priestley but not Lavoisier. Clearly, the appearance of each item on anyone's list is embedded in decisions which are less than self evident. In our earlier discussion of the discovery of oxygen, the citation of Priestley and Lavoisier takes considerable licence with the historical materials, ignoring the arcane character of both Priestley's phlogiston and Lavoisier's 'principle of acidity', and reconstructing the histories in light of the later developments in nineteenth-century chemistry. Likewise, the controversy between Bateson and the biometricians at the turn of the century led Bateson and his followers to reconstruct Mendel as a corroborator of Bateson's model of discontinuous variation, in spite of the fact that Mendel's work constituted a theory of hybridization as a clue to speciation, as opposed to hybridization as a clue to genetic variability. All of the judgements and social processes which go into the production of a list of multiples are effaced by the simple serial identification of the discoveries and their claimants. In other words, the collection of several researchers under the title 'discovery of oxygen' or 'discovery of genetics' presents an image of uniformity that is hardly warranted.

As Kroeber indicated in 1917, a systematic exposition of each citation would certainly fill a large volume. However, the present remarks, though they do not investigate each entry, tend to challenge the validity of the 1922 list. Also, our observations of the judgemental context of discovery suggests that a theory of discovery should not focus on how ideas come into the mind, or how they evolve as the culture matures, but how they are *defined* as discoveries. Obviously, the judgement whether something is a discovery or not depends on such things as the perception that a new idea was not 'borrowed' from someone else, and that the idea was not 'old hat' and preceded by earlier equivalent work. In other words, many of the decisions which went into the preparation of the list of multiples are elements which contribute to the common sense perceptions of discovery. This model of the problem confirms that we should consider not what makes discoveries happen, but what makes them discoveries. Besides the culturological model which we have just examined, the great proportion of writing as outlined in earlier chapters has been devoted to the identification and study of *psychological* models of discovery. These have uniformly attempted to account for discovery by explaining how ideas get into the mind. A principal explanatory 'force' identified here has been called genius or some equivalent. This is the second folk theory which I would like to discuss.

GENIUS: THE SECOND THEORY OF DISCOVERY

Though very few modern writers explicitly rely on the position of Galton in *Hereditary Genius,* this is not because they do not think that genius is a crucial factor in the production of discoveries.[24] Rather, it is because they do not have strong reasons for believing that it is hereditary. Nonetheless, in the wave of writing on discovery which has appeared in the last two decades, references to genius occur with regularity. What is of note is that such references are often ad hoc. They are offered typically when the principal explanation falls short of the mark in empirical cases.

For example, Kuhn's model of discovery relies on the concept of the paradigmatic order of normal science and the anomalies which become associated with it. According to Kuhn, discoveries occur with the recognition that nature has violated the expectations created by the paradigm. However, not all novel facts require the reformulation of the model of nature. Kuhn distinguishes between mere novelties of fact and theoretical discoveries. While the discovery of Uranus was a simple factual discovery, the discovery of oxygen was a theoretical discovery for it entailed a theory of combustion, molecular weight, the periodical table, etc. But how is one to recognize whether a particular unprecedented observation is a mere novelty, as opposed to an anomaly? In the discussion of the discovery of oxygen, Kuhn notes that various researchers like Stephen Hales isolated pure samples of oxygen decades before Lavoisier. However, Kuhn suggests that the gas was merely novel for Hales, and that one of the 'normal requisites for the beginning of an episode of discovery . . . is the individual skill, wit or genius to recognize that something has gone wrong in ways which might prove consequential'.[25]

In retrospect we have no problem identifying those novelties which turned out to be significant discoveries, that is which were truly anomalous. However, in a contemporary perspective the only measure Kuhn would have of an anomaly is whether the researcher has the skill, wit or genius to predict a discovery: consequently, no discovery, no genius!

One finds the same kind of circularity in the writings of Hanson. Hanson suggests that discoveries are made by the mental process of 'retroduction' – retrospective reconstruction. This procedure suggests that, even as in finished theories there is an inference network from high level generalizations down to low level descriptions, so too there must be a reverse process in the inference network whereby a scientist constitutes a set of high level constructs to account for certain previously witnessed problematic (anomalous) observations.[26] How is this done? Retroduction! However, the only evidence of this ability is the very conclusions which it is purported to have brought about. Consequently, it is hardly surprising to find Hanson[27] reflecting with circularity on the fact that perhaps the basic laws of motion were so profound that they required the very great intellects of men like Galileo, Newton and Kepler to fashion them. Without geniuses like these, the laws of physics would have remained undiscovered. Yet the only evidence of this genius is the success which it produces.

Richard Blackwell[28] argues similarly when he compares the success of Newton to the failure of Descartes regarding the law of free fall. When Blackwell's analysis cannot account for the difference, he suggests that genius must play a part in such matters, for how else can we explain the different performances?

The etymological roots of 'genius' point to the notion of an internal guiding spirit, something akin to a secular soul. Consequently, it is hardly surprising to find analogues to the notion of genius in 'the subconscious', which also refers to an internal guiding spirit or 'genie'. The theories of Koestler[29] and Poincaré[30] rely explicitly on models of subconscious guidance. For Koestler, discoveries come about as a result of the subconscious synthesis of ideas and their application to outstanding questions. So too, Poincaré argued that the subconscious was an almost *infallible* source of insight. For both Poincaré and Koestler, discoveries occur as 'flashes' of realization which boil over from the subterranean mind and intrude into the consciousness of everyday life. We shall discuss such 'gestalt switches' shortly. In terms of the present discussion it is enough to note again that the mechanism employed to explain the appearances of discoveries is tautological or circular. The synthesis which the subconscious produces *is* the discovery: it does not explain it. This has been a perennial problem with gestalt accounts – the change in consciousness is the very thing that constitutes the discovery. That a discovery occurred in or as a gestalt shift tells us *that* a discovery has been made. It does not explain what produced it.

In summary, we observe that among all the prominent writers on discovery, there is a recurrent reliance on the mentalistic concept of 'genius' or some equivalent. These models make the problem of discovery equivalent to the question of how ideas come into the mind. Furthermore, the accounts of this process are typically circular or tautological, and are often offered in an ad hoc fashion when a preferred account cannot handle a particular example. While the usual standards of inquiry suggest that we abandon such models and seek others, the present investigation finds these 'theories' of interest in themselves, for in spite of the fact that they are illogical, *their recurrence is so ubiquitous and their appeal so common sensical as to recommend them as pragmatic accounts used by members of society to make sense of discovery. In other words, we should treat them as common sense theories or folk explanations* which no matter how problematic for the methodologist, are natural and compelling for the member of society.

Discovery and reciprocity of perspectives

When considered as folk accounts, what force do these theories have as explanations for members of society? What is there about discovery which is especially problematic, or requires special accounting for, and for which the notion of genius is specifically adequate? The answer lies with the 'mundaneity principle' developed by Melvin Pollner[31] and the reciprocity of perspectives discussed by Schutz.[32] The central premise of Pollner's mundaneity principle is that in everyday life we typically assume that the world is objectively structured and that its elements are independent of our own construction of them. Similarly we assume that were we to switch perspectives with our fellow beings, we would experience the same worldly events from their point of view. Consequently, when we are confronted in everyday life with inconsistent accounts of a phenomenon, this reflects the perspective of one of the observers, and not the structure of the real world. When such inconsistencies occurred in the traffic court he studied, Pollner noted that the judge handled these inconsistencies by challenging the integrity of one of the parties, or showing that the inconsistencies were superficial, that is, that the witnesses were talking about different events.[33] Hence such devices resolved the potentially problematic inconsistencies in the reciprocity of perspectives and the apparent challenge to the mundaneity principle.

With discovery we have an analogous problem. In the natural attitude we assume that the world is objectively structured and independent of our accounts of it. However, in such a domain how can we grasp the fact that someone may come to apprehend parts of that world which are not yet known, yet which have this knowable-in-common quality? If the natural world is knowable-in-common, but is not concretely known in all its particulars, the identification of further aspects of the world, like new laws, puts some strain on the reciprocity of perspectives. How is it that someone else came to apprehend the new law?

The issue here is not resolved with claims about a division of labour which hold that someone made a discovery because he was doing research, as though making a discovery were like building a house. Unlike building a house, making a discovery is unique. That is, the procedures adequate for the job are by definition not part of the public domain, whereas for building a house they are. So it is not a matter which is resolved by claiming that the perspectives of the discoverer versus the member at large are concerned with *different* events. Recall that there are typically several researchers directing their efforts towards the explanation of the same events. When we find such a situation, and when we find that only one scientist has succeeded, especially when there is evidence of rivalry, the idea that the discovery was a 'stroke of genius' on the part of the successful candidate is an attribution of a *power* to the perspective of that discoverer which explains the difference in his or her perspective from both other researchers and the member of society at large. In other words, *by making the perspective of the discoverer extraordinary, the mundane assumption of the reciprocity of perspectives is preserved by admitting of special exception.* We assume a reciprocity of perspectives with other members of society. However, there is a prevalent notion that the genius 'lives in a world of his own'. Like Einstein, he is characteristically eccentric. Handlin[34] has traced this ambivalence to genius in his study of the literary images of Doctors Frankenstein, Jekyll and Faust. The same asymmetry is conveyed by dystopian novels like *Brave New World* where scenes of technological rationality evoke feelings of anomie and tragedy. At least in these literary works, the genius and his work are *not* sources of a reciprocity of perspectives. He is depicted literally as the stranger in a strange land.

Also by assuming that the discovery occurred as a function of genius, we provide for the fact that the discovery is no mean feat, but is a profound achievement, 'an act of genius'. Consequently, the more fundamental the law, the greater the evidence of the genius. As Hanson notes, the basic laws of motion were so extraordinary that only a few people had intellects mighty enough to uncover them. On the other hand, by assuming that the mental prowess of certain scientists is awesome, the occurrence of the discovery is made highly accountable. That scientific laws have been uncovered is to be *expected,* given the profound mental abilities of successful, scientific researchers. This is all folk reasoning. In other words, for the sociologist the common sense theory of genius makes the appearance of discoveries unproblematic, in spite of the otherwise assumed reciprocity of perspectives. For the folk member of society, it provides for the fact that he or she did *not* make the discovery, while explaining why the discovery *was* made by some particular and unique individual. Furthermore, the assumption that genius is a mental power or force strengthens the naturalistic sense of discovery which assumes that discovery is an objective 'effect', which is made to occur by an independent variable which appears earlier. Consequently, common sense not only provides the method whereby an event is produced as a discovery, but provides in the associated formulation of genius the method by which the local identity of discovery is seen to be an independent creation, as opposed to a local act of construction.

The explanation of discovery from cultural maturation has this same feature. If, then, discoveries are seen to occur inevitably 'with the march of history', their occurrence for members of society appears to be independent of local production. Merton[35] notes that the argument of multiple independent discoveries, far from being the prerogative of twentieth-century anthropology, is 'self-exemplifying'; it has appeared repeatedly in numerous commentaries on science. Merton mentions among others Benjamin Franklin, Lord Macaulay, Auguste Comte, Augustus de Morgan, Sir David Brewster, Samuel Smiles, François Arago, Frederick Engels, François Mentre, Pierre Duhem, Emil du Bois-Raymond, George Sarton, A. L. Kroeber, Albert Einstein, Nicolai Bukharin, and lastly, Ogburn and Thomas. The repeated use of such observations to recommend the argument of cultural maturation, though logically erroneous, indicates its function as a folk explanation. However, it accommodates itself to the mundaneity principle in a different way. Instead of providing the special circumstances under which the reciprocity of perspectives can be deemed accountably irrelevant (that is, in cases of genius), the cultural argument anonymizes the question of how discovery arises. Discovery appears in a Durkheimian fashion, as though it were not particularly the object of individual, human agency. In fact, the cultural view understands discovery as though its appearance were *inherent* in the progress of society. Consequently, in this account there is no strain on the reciprocity of perspectives; in fact discovery seems to be a natural fact par excellence, a kind of exemplary object which makes the reciprocity of perspectives possible by providing for the world a natural objectivity independent of human action.

In summary, the explanations of discovery based on genius and cultural maturation, though inadequate in strict scientific terms, have occurred so repeatedly and contain such strong intuitive appeal as to suggest that they are elaborate common sense beliefs. They provide an objectivist interpretation of discovery which both preserves the assumed mundaneity principle which Pollner associated with the reciprocity of perspectives, and corroborates the naturalistic assumption that discoveries are brought about by exogenous forces. The occasioned use of such an account by members of society is reflexive; it provides an apparently independent set of referents of the usage (genius or cultural maturity) which is proof of its relevance and force, and only available by taking the usage for granted. Specifically the usage of genius accounts for the fact *that* the discovery was made, that it was made *when* it was made, that *who* made the discovery was not accidental, and that its status as a discovery was an objective and naturally occurring fact. The reflexive character of these common sense theories has been treated as logically circular, or tautological, in chapter 2. However, this circularity is a consequence of the reflexive quality of the collectivity member's accounts, which are inherent aspects of the settings they elaborate.[36] That circularity is a logical error for the naturalistic scientist; however, for the sociologist, it is a crucial observation in the study of folk phenomena.[37] There is one further issue. The study of folk elements in a familiar culture must not only show how members of society 'produce' their worlds, but must focus on how the members of society are typically unaware of their own role in reifying social facts. We have discussed reflexivity in this regard. However, there is a further clue to this process in another folk theory, namely, gestalt shift.

GESTALT SHIFT AND THE EXPERIENCE OF INADVERTENCE

If, as James Heap[38] has suggested, the common usage and the technical usage of linguistic expressions constitute a continuum of significance, then the mundane usage of discovery further corroborates the reflexivity of its folk use. *Central here is the sense of inadvertence associated with the mundane usage of discovery.* This provides another ground for the member of society's sense that discovery is independent of his local production. Consider these sentences: 'After he had locked the car, he discovered that he had left the keys inside.' 'On return from a tour of peace keeping duty in Cyprus, a young army officer discovered that he had contracted syphilis'. In these cases the recognition of discovery is guided by the perception that these states of affairs were factual, and that the statements describing them were true. Also, these facts were new and emergent; at one time these things had not yet happened, but only occurred subsequently. These attributes are common with the scientific sense of discovery. However, it is clear that the mundane sense of discovery is not by definition relevant to science. The discovery that one had locked oneself out of one's car, or had gotten oneself infected, would not qualify one for a Nobel Prize. These are not discoveries of natural laws, or phenomena in the scientific domain. Also, although these events are sensible in terms of our common stock of knowledge,[39] that is, things we account for with our knowledge about the structure of locks and the consequences of lovemaking, their occurrence is not provided for specifically. Nor have they occurred in the course of scientific research. This tempers the sense of discovery considerably.

In science the unexpected or the unknown is sought after. Scientific discovery is 'the expected unexpected'. By contrast the common sense use of discovery refers to the patently *inadvertent or unanticipated* experience of the act. Consequently, the sense of possibility in science shades the meaning of uniqueness or novelty. Certainly scientific discoveries are news; however, it is clear that they occur in the context of courses of action which are motivated for their disclosure. When we say that so and so discovered that he locked himself out, it is clear that this event was not the outcome of a motivated course of action. It occurred unexpectedly 'because of' a contingent inadvertence, whereas scientific discoveries happen in the midst of actions pursued 'in order to' reveal them.[40] However, there seems to be a common foundation to the two usages: *both refer to a topical revolution in one's experience* or stream of consciousness. This account illustrates the point.

I had left the store and was walking to the parking lot thinking about how rude the clerk had been and imagining how I might have come off better in the argument than I actually had, if only I'd insisted on seeing the manager and giving him a . . . then it occurs to me that something is amiss and I discover I've left the baby sleeping in the carriage at the checkout counter. I think of nothing else. All recollections of the argument disappear as I rush back immediately with pangs of anxiety.

Similarly as I complete the act of getting out of the car, pushing the handle to 'lock', and slamming the door shut, my eyes catch sight of the keys still in the ignition. I freeze immediately. How shall I get back in? The sequence is disrupted. These thematic shifts in experience are the kinds of events which we designate in everyday life with the word 'discovery'. These discoveries constitute immediate reorganizations of the matters at hand; they entail immediate prospective and retrospective considerations: how shall I get back into the car? Or, where did I leave the carriage? I must return to the baby! Did it happen on the tour of duty? Who was the contact? What shall I tell my wife? etc. All these constitute more than the emotional element of *surprise* which Koestler[41] discusses in terms of emotional release; they comprise a compacted reorganization of one's relationship to others and/or the world which displaces the sequence of action by bringing to consciousness others which were more compelling. For example, Archimedes foregoes his bathing in favour of his scientific research; he has reconstructed the crown of Hiero as an object of hydrostatic displacement. The fact that the thought appeared in the manner in which it did lends to it its *involuntary sense of occurrence*. Likewise, Gutenburg sees beyond the festivities of the wine fest and the wine press and views the printing machine under an analogous 'press'. In all these cases, the discoveries appear as intrusion of consciousness; they disrupt the stream of activity and plainly, under such circumstances, it can hardly be said that they are made to appear. This patently involuntary structure of one's thoughts makes it plain to see *for members of society* that discoveries are independent of their *in situ* production *by members of society*. In other words, *the inadvertence which members of society associate with discoveries* especially mundane discoveries from their own experience as well as those in science which like Archimedes' confirm this, *make such discoveries appear as objective or natural incidents*. Consequently, this sense of inadvertence, like the common sense theories of genius and cultural maturation, only adds strength to the member's recognition of the naturalistic foundations of discovery.

However, for the sociologist the inadvertence which members of society experience and account for as 'gestalt shifts' or 'flashes of insight' are nothing more than situationally contingent changes in our awareness of topics. Gestalt switches are in fact *integrative* events; typically a sequence of thematically unimportant action is displayed by a *pre*occupation of greater interest. For example, Poincaré's flash of insight regarding the nature of Fuchsian functions[42] displaces the routine of boarding the bus. The man who inadvertently locks himself out of the car now finds he has a real problem on his hands, as does the infected officer.

In the experience of thinking or reflection our consciousness not only wanders but leaps from topic to topic, from the world of the lecture to the world outside the windows, from pragmatic matters at hand, to personal interests and back again. This is an inveterate characteristic of conscious life and has been expressed metaphorically as 'the flux of experience' and 'the stream of consciousness'. Such being the case, scientific discoveries *may* occur as thematic shifts in topical awareness; they may constitute contingent surprises or inadvertencies. However, their sensibility as discoveries is given by a prior awareness of the problem. In the same vein, an officer who has avoided casual sexual intercourse is not likely to interpret symptoms of bladder irritation in the same manner as his more sexually active friends. If for example over breakfast one of the latter should discover or realize he has the symptoms, then the lovemaking, the discomfort and the inflammation, the burning sensation and the sense of defilement all crystallize forcefully as a unitary sequence of activity that displaces the routine of eating and holds all other thoughts at bay.

That this awareness 'forces' its way into consciousness, superseding reflections, only testifies to its importance. It is hardly an enigmatic problem that scientists should experience shifts in awareness in such a manner, and that some such shifts should be experienced as remarkably important; science is after all one of the preoccupations of the scientist. For the sociologist, that such shifts are 'inadvertent' is a characteristic of the stream of consciousness; however, for the member of society such shifts may be experienced as evidence for the naturalistic basis of discovery. The latter conclusion emerges from a confusion of the ostensively inadvertent *timing and placement* of the occurrence of the event, with the *significance* of the event to consciousness. In other words, because an idea appears 'to come out of the blue', it appears to be naturally occurring and therefore appears to require a naturalistic explanation; for the sociologist, the placement may be inadvertent, that is, situationally contingent, but it is already given within a horizon of significance by which it is illuminated. Hence the local sense of inadvertence, like the recurrent accounts of genius and cultural maturity, strengthens the familiar sense of discovery's exogenous origins, and hence these things lend a sense of independence and objectivity to what are nonetheless for sociology judgemental or attributional statuses.

By way of summary, let me suggest that, in contrast to the model advocated in this study, all the prominent accounts of discovery, whether found in anthropology or psychology, appear to adopt a naturalistic model of the problem in which discovery is treated as an 'effect' of an exogenous 'cause'. These theories have been of two types: accounts which explain discovery by pointing externally to cultural maturation, and accounts which point internally to some 'power' within the individual. What is of note chiefly is not that these theories are inadequate, that is, that they are tautological, or empirically unsupportable. Indeed, there certainly have been cases of simultaneous discovery which have been neither vastly different in quality nor subject to plagiarism. And scientists are by and large relatively clever men and women. However, whether these observations support a social or psychological determinism which accounts for the appearance of discoveries in society is another matter. What is interesting is that despite the obvious logical and empirical shortcomings of these positions, they occur with uncanny regularity in this society and constitute some of the chief ways in which discoveries are objectified. The objectification of discoveries as acts of genius provides an ad hoc solution to the usual assumption of the reciprocity of perspectives; similarly the idea of cultural maturation, with its implicit teleology, makes the discovery seem inevitable, and lends a sense of transcendence and objectivity to the achievement, which makes its disclosure by any individual seem natural and unavoidable. The experience of gestalt switch which is reported widely in this literature heightens the causal sense of appearance found in these two theories. Unlike the previous writing which tries to outline the social bases underlying the 'occurrence' of particular discoveries, this chapter has focused on theories *about* such processes to uncover some of the strategic beliefs about discovery which, because of their familiarity, obviousness and common sense, are for the investigator typically invisible.

9

Some sociological features of discovery

For the member of society, the phenomena of discovery are *substantive:* they encompass the ostensive content of the discovery, and naturalistic understandings of how the discovery originated. By contrast for the sociologist, the phenomenon is not the naturalistic status of the discovery *per se,* but the common sense *use* of discovery as a method of identifying and elaborating scientific events. *What* that method consists of for members of society has been discussed in terms of criteria of intelligibility which constitute the recognition of discovery.[1] However, the question of *how* the status of the events as discoveries is maintained and evolved has been given little attention. The question of *how* has to do with *basic* interpretive practices which underlie the criteria and organize their use.[2] Employing Chomsky's metaphors, Aaron Cicourel has advised that we should treat the member of society's phenomena as 'surface' rules which are maintained by a 'deep structure' or 'machinery' of interpretive practices.[3] In this case, the surface or common sense structures are the sense of unprecedentedness, validity, scientific relevance and substantive possibility. These constitutive elements are grounded in and by the interpretive practices as illustrated by the most basic of such practices, the documentary method of interpretation.[4] This procedure consists of Ego's interpreting actually witnessed behaviours as exhibits of an assumed underlying pattern or a normal form of the state of affairs.[5] Similarly, 'other things being equal', Ego's perspective on events is presumed to be the reciprocal of Alter's perspective, where the differences in what they report is only a reflection of the limited vantage point of the observer, and not in the phenomenon observed. This assumption of the reciprocity of perspectives is central to the main thesis of mundane reasoning developed by Pollner, and discussed earlier. Also the sense of the surface structures may become modified over time because each successive datum identified by the documentary method may modify the whole assumption of the interpretation as it proceeds, providing a different sense to past events and a changing sense to future events.

This may involve the retrospective reconstruction of meaning, as well as changing prospective expectations about the as of yet unwitnessed particulars. These sense making practices – the documentary method of interpretation, the assumption of normal forms, the reciprocity of perspectives, and the retrospective and prospective constructions of appearances – all operate to gloss or highlight certain actual details of a setting so as to make apparent the relevant common sense elements of discovery which the setting is assumed to contain.

The present study has paid more attention to the fact *that* social understandings are central to the status of scientific discoveries; this emphasis has been derived from the need to break the grip of a naturalistic sense of discovery prominent in other writings. Consequently we have become vividly aware *that* discoveries are methods, at the expense of understanding *how* such methods have operated. Also, since this study is confined to written accounts of discoveries, as opposed to *in situ* investigations of scientific research, the disclosure here of how these methods operate is only an appreciation as opposed to a uniquely adequate rendition of the operation of such methods in actual settings.[6] Nonetheless we have had some striking illustrations even from the documentary evidence.

For example, it was noted that when the Piltdown find was first announced, it was identified by those present as 'the missing link' about which Darwin had speculated. The fact that Darwin had not specified that an *'Eoanthropus'* would have simian jaws combined with a developed skull was 'allowed to pass'. The Piltdown remains were 'more or less' what one would expect of a missing link. On the one hand, those like Waterston who disputed Dawson's interpretation of the remains, and who held that *two* separate creatures were involved, pointed to the patent inconsistencies of the bones; how could such a 'thick' mandible be organically associated with such a 'normal' skull? Also, since the back of the jaw was missing, was it not impossible to show exactly how the two fitted together? On the other hand, were not all the remains identically coloured? And were not the associated bones of large mammals very old, and the tools very primitive? Did this not make the *Eoanthropus* exceedingly old? All these questions were directed, and in part motivated, by the particular assumption that was brought to bear on the remains. If one assumed like Dawson that the remains were homogeneous, this underlying assumption led one to order all the details as further documents identified in light of this assumption and in fact further corroboration of it. The assumption of two creatures also ordered the remains as a corpus of similarly supporting documents. The value of any detail therefore was indexed to the underlying assumption. The assumption led one to infer from the remains whatever course of action was deemed evident to produce remains characterized by an accidental contiguity on the one hand, or a necessary coherence on the other.

But how could the scientists assume the thesis of mundane reasoning while holding such mutually exclusive interpretations? Both Weiner and Millar note that the controversy of the announcement resulted in a 'wait and see' posture. In other words the remains took on a prospective status. If future disclosures could not have corroborated Piltdown Man, then the initial remains would have been reconstructed in a more dubious light. However, when the eyetooth was recovered, and when the remains of a second Piltdown skeleton were reported, the two-creature theory appeared untenable, and the coherence and antiquity of *Eoanthropus* appeared well founded. Consequently the remains were seen to have a 'normal form' or a typical organization in that the peculiar association in the initial display now appeared to be standard for those remains;[7] furthermore, that form was now seen to be a fact for anyone's inspection so that any displays of the non-reciprocity of perspectives could be chalked up to personal stubbornness, bias or incompetence – all of which challenged the credibility of the witness in favour of the integrity of the discovery.

Another aspect of the reciprocity of perspectives was that, in spite of their inconsistent judgements, scientists like Keith and Woodward each assumed that the other held the remains to be the results of sincere scientific research. When the identity of the event was reconstructed by Weiner to cohere with the evidence of artificial manipulations, that is, when Weiner recognized the forgery, this reciprocity of perspectives regarding the original announcement was reconstructed. There was no reciprocity – but only a *pretence* of reciprocity on Dawson's part. On the other hand, Millar recognized the hoax, and the pretence is ascribed to G. E. Smith. In any case, the reciprocity of perspectives was an assumption in force for the recognition that Piltdown was an actual scientific achievement, and was one of the dimensions which was modified in the course of later rehabilitations of the remains. However, even in these rehabilitations, we assume a reciprocity of perspectives between ourselves, and either Dawson or Smith, depending on whether we follow Weiner's, Millar's, or Douglas' reconstruction. We assume that we see the fraud or the hoax, just as (either) Woodward or Smith would have seen it and as if in fact it had been a fraud or a hoax all along.

Indeed, when a claim of forgery is made, the events are reconstructed along the 'normal' or 'characteristic' form of such forgeries.

Weiner characterizes Dawson as something of a dilettante motivated by personal aggrandizement, and not surprisingly he relates that Dawson had perpetrated other forgeries.[8] He notes specifically that Dawson's two-volume work on Hastings Castle was copied from an earlier unpublished manuscript written in 1824 by William Herbert. Also he relates a report from an 'eye witness' who claims to have 'caught' Dawson staining the fossils; Dawson is characterized as reacting with the typical shame and embarrassment of one caught 'red-handed'. These are the assumed normal forms of a case of forgery: the act is only one episode in a series of deceits and the surprised forger exhibits the telltale signs of guilt. On the other hand, Millar reports that Dawson freely admitted using Herbert's manuscript, and that Dawson even explained how he augmented it with Herbert's later drafts.[9] In fact two-thirds of his introduction is a fulsome acknowledgement of his indebtedness to Herbert who had forestalled Dawson's own investigation of the castle. Also regarding the episode of staining, Millar points out that it was widely though erroneously believed at the turn of the century that fossils were *preserved* by chromium staining. If Dawson was doing this, the reason is apparent: the 'eye witness' had entered Dawson's laboratory without waiting for him to open the door; that is, Dawson was startled, not embarrassed.

Clearly the interpretations given to various episodes in the Piltdown record are managed by different uses of the documentary method of interpretation, different assumptions of what events were perceived reciprocally by the participants, the reconstructions of earlier details as the case was unfolded, and the prospective expectations about what other matters could be relevant and what they could be expected to reveal. Such practices made the formulations about the discoveries accountable, commendable, as well as contestable issues for participants in the matter.

They were also evident in the various approaches to the Mendel paper, read at one time as normal breeders' work (by contemporaries), then as proof of discontinuous evolution (by Bateson and De Vries), then as a suspicious manipulation of statistics (by Fisher), and finally as a systematic collection of earlier findings (by Zirkle). Each interpretation was produced through the use of a different underlying assumption by which all the relevant circumstances of the document were seen to fall into place. In fact, each successive investigation constituted a *reconstruction* of the paper by placing it in a different set of relevancies.

Similarly the predecessors of Columbus who 'discovered America', and the discoveries of oxygen, are reconstructed from hindsight in such a way that the local form of their achievements is edited to resemble contemporary understandings. Specifically Lavoisier's arcane principle of acidity is played down, while oxygen's tie to combustion theory is emphasized. Also all the pre-modern explorations of the western hemisphere are reconstructed as discoveries of 'America', and Columbus' own efforts are read in the light of Vespucci's contribution. In these cases the local sense of the events is reordered for us by the circumstances which postdated them – so that the characteristics of those events is assumed in retrospect to be equivalent to their current form.

In summary then, the approach developed in this investigation has clarified the sort of phenomenon which discovery is, what that phenomenon consists of, and how it operates in actual settings. Specifically it has been argued that discovery is an attribution of a social status conferred according to criteria of intelligibility which operate collectively as a method of interpretation used by members of society to organize the world of scientific achievements. That method consists in the use of basic sense making practices to establish the meaning of an event in terms of the constitutive elements of discovery. These defining elements constitute a common sense knowledge of discovery and consequently allow the member of society to determine what can be recommended as a discovery and who can be recommended as a discoverer. Furthermore, these are the things which, when perceived, forcefully document the apparent objective status of the event.

A leading aspect of the approach we have pursued is that the status of discovery is not conceived of naturalistically nor are discoveries seen to occur as *effects* or *determinations* of exogenous variables. By contrast the occurrence of discovery is a locally accomplished definition attributed to an event by competent members of society. However, for reasons I described in chapter 5, the criteria nonetheless do constitute a conditional form of explanation. Under conditions where members of society perceive these elements, they will by 'force of reason' be moved to conclude that a discovery has occurred, and without these conditions, they will attribute a different status to the event. Simply put, what are the conditions of discovery? Criteria of intelligibility. How do they explain discovery? By showing the competence of members of society to define events appropriately. For the sociologist then, discovery is a practice or social process which objectifies experience, making it appear to members of society to be naturalistic and/or independent. That sense of independence is accomplished through the identification of a set of 'corroborating' circumstances or details that co-emerge with, though are mutually dependent on, the assumption of the discovery. Also the folk explanations of discovery as well as the sense of inadvertence in commonplace discovery discussed in the last chapter aid the naturalistic interpretation. Hence discoveries *per se,* and all the associated characteristics which they have for members of society, are socially constructed by the endogenous operation of the interpretive practices of members of the culture.

Naturalistic approaches have overlooked the *performative* usage of discovery in local settings.[10] By asking 'what made the discovery occur' as opposed to 'for whom and how is some event an occurrence of discovery', a naturalistic approach tends to miss the local discussions of the achievement, wherein terms like 'discovery' or 'breakthrough' or even 'forgery' or 'plagiarism' have a *signal function*.[11] For methodological reasons, the signal function of these utterances is not, for the sociologist, proof that a discovery has occurred, but evidence of the local commitment of the participants to that interpretation of the event. In other words, local discussion about a discovery should raise questions about how the discovery was *decided,* as opposed to how it arose from nature or from the scientist's subconscious. On the contrary, naturalistic approaches tend to operate on the assumption that the discovery status is an internal attribute of its identity. And in fact, the question about the social derivation of the identity appears to be precluded by the imperative in naturalistic methodologies to define terms strictly as arithmetic quantities rendered in orthogonal functions. In other words, the strategy of re-constituting social phenomena as 'variables' substitutes for their local meaning-in-use the analyst's formal definition which is typically limited to quantitative attributes; the placement of the objectified term in a hypothetical 'relationship' to other terms adumbrates the quasi-performative nature of its local usage, and the judgemental or motivational relationships its local usage signals. In other words, the practice of ignoring the local status of discoveries and searching for their underlying or external origins tends to further mystify discovery, by attributing it to independent exogenous conditions, at the expense of the endogenous reasoning. We have tried to avoid this.

Exogenous and endogenous accounts

The formulations of the origin of discovery which we examined in the first chapters of this work specified a number of exogenous independent variables which were said to bring about discovery. These were found to be logically erroneous and/or empirically inadequate. However, in the context of the present explanation, they are seen as a species of folk reasoning used by members of society to account for how discoveries occur. As such they are aspects of the endogenous origin of discovery which we have been examining in these last chapters.

The endogenous account abandons the causal or naturalistic form of explanation favoured by other authors. Discoveries occur because they are made to occur socially by processes of social recognition. There were indications of such processes in the troubles associated with Thomas' construction of the list of multiple discoveries; whether an event was listed as a discovery or not was emphatically a matter of *judgement.* That judgement involved reconstructing the conditions under which some achievement was made, the local success of the discovery, the integrity of the research and the researcher, the relative placement of the announcement both in the community and in the tradition, etc. Clearly the status of the event as a discovery was not naturalistic, but was socially mediated. The same issue emerged in Merton's work on priority disputes. These were not simply matters of *who came first,* but *which* achievements were in fact discoveries, and which were duplications or plagiarisms. Again this pointed to the social evaluation of the achievement. Discovery turned out to be a particular kind of evaluation with certain inherent features – relevance, possibility, validity and uniqueness. The discoveries which we examined in detail made this all the more evident. The recognition of discoveries by members of society involved the attribution of these features as a method for elucidating the organization of the event as a discovery. The radical changes in status of Mendel and Dawson vividly underline the social processes by which the 'objective' status of discovery is managed. This further corroborates the view that for sociologists scientific discoveries are folk phenomena which, like Azande magic, are used by members of society to make aspects of the environment sensible or accountable.

The endogenous use of discovery as a scheme of interpretation is rarely transparent to members of society. In this study, in order to make the social production of discovery available as a folk phenomenon, we have examined a series of discoveries whose statuses have been problematic. However, members of society have typically missed the reflexive and endogenous character of their interpretation of discovery because the very phenomena at stake are re-ordered by each successive interpretation. For example, the recognition that the Piltdown remains were fabricated displaces the earlier sense of discovery by relegating the artifacts to acts of forgery or hoax. Hence the event is *not* relativized by successive interpretations but retrospectively co-opted or re-constituted as having been something else all along. Similarly since the time of Columbus the earlier non-European explorations of the western hemisphere have been interpreted as *forerunners* in the discovery of America. It is evident that the use of such interpretations re-orders the significance of the phenomenon so that the relativizing potential of competing definitions of the situation is dismissed. Also, the principle of mundane reasoning assures to the member of society the sense that the world is a singular, natural and objective environment. Consequently it makes discovery *un*available as a relative, endogenously specific accomplishment. This occurs because each interpretation provides and is equated with a set of coherent elements that not only explain a phenomenon, but do so in a way which co-opts and accounts for competing interpretations. These elements constitute apparently independent evidence of the discovery, while at the same time relying on the sense of discovery for their relevance. For example, it is significant that all the profound and compelling details which supported the Piltdown *discovery* became irrelevant and are quickly forgotten when the sense of *forgery* or *hoax* becomes prominent. The new corpus of supporting evidence is also animated by a specific interpretation (of hoax or whatever) which generates its own corpus of elements and appears to be independent and self-evident.

However, the cohort independence has been guaranteed in other ways. The ubiquity and intuitive appeal of the arguments based on genius and cultural maturation, no matter how logically and empirically deficient, nonetheless function as naturally theoretic accounts of discovery. These accounts provide a further sense of the discovery's independent status and origin by providing for its exogenous basis. In other words, these accounts foster the conviction that discoveries result from natural causes.

The mundane sense of discovery which is characterized by the experience of inadvertence functions in a similar fashion. This situational inadvertence of insights and realizations provides a vivid proof that discoveries occur 'involuntarily', which again supports the conviction that discoveries happen as natural phenomena, independently of the beliefs of society. All such practices recommend the folk sense of discovery as a naturally occurring phenomenon, which requires naturalistic explanation. However, none of these practices can be taken *by the sociologist* as proof that discoveries have 'actually' occurred, or that they in fact require an 'objectivist' explanation. The orientation of members of society to the novelty, validity, relevance and possibility of a scientific achievement is not of interest as 'real evidence' of a discovery, but as an endogenous folk phenomenon or practice by which such things as discoveries are locally recognized as compelling social facts.

In other words, the natural sense of discovery for sociology is a folk phenomenon. However, the sociological identification of this is itself an achievement which the present work recommends to sociology as its own discovery, with the conviction that this work is part of the very order which it describes.

THE PROGRAMME

The remarks which constitute the theory in this study do not contain an exhaustive open and shut case for a model of discovery, but describe a whole programme of research with implications for past, present and future investigations. We have focused our attention initially on the alternative models found prominently in the current literature, and while showing the costs which belief in these incur for the analyst, we have also explored their folk or common sense value. In contrast to these various naturalistic models, whether found in psychology, anthropology or everyday life, our alternative theory has been outlined. While it is not necessary to recapitulate the entire perspective, suffice it to say that this theory, by virtue of its reliance on the criteria of intelligibility of discovery, has described a model which is both meaningfully appropriate and causally adequate. It is causally adequate in that the criteria constitute a set of individually necessary and collectively sufficient conditions for the recognition of discovery; and it is meaningfully appropriate in that the criteria are the elements which members of society themselves recognize when they interpret events as discoveries. This model has allowed us to rehabilitate the common, though erroneous, understanding of the place of Mendel in the history of biology, and to substitute our own more compelling model for a host of unconvincing explanations for his long neglect.

We have also examined the utility of this approach in a discussion of certain other discoveries involving America, oxygen and Piltdown Man. In the context of this discussion, a number of issues were raised regarding the objectification of discoveries, and their transformations over time. These observations will, I hope, be corroborated by the intensive study of future cases. However, this approach has consequences for further studies in related areas. Several types of investigation which will consolidate and extend the present approach immediately come to mind.

On multiples

First of all, this approach presents substantive reasons for re-evaluating the lists of multiple discoveries. In other words, the book

Kroeber called for in 1917 should still be written. I would conjecture that a large number of entries in the Ogburn and Thomas list could easily be revised, as we have revised the entries for the discoveries of oxygen and genetics. However, though this does not suggest that multiples do not occur, it does suggest that there are probably a large number of things which control their occurrence in addition to the 'maturation' of culture. I would conjecture that, for example, if we inspected the earlier entries, as opposed to contemporary ones, we would find strong evidence for the effectiveness of shared research traditions replacing the ineffectiveness of publications as conditions of importance changing over time. Nonetheless, I think we would still have grounds for believing that, even in some of the best cases, the identity of the contributions would be matters of dispute reflecting the interests of particular intellectual communities, and the performative aspects of their adoption.

Recent work by Edward W. Constant II strongly supports the position on multiples discussed here. Notes Constant, 'Conventional interpretations of multiple invention depend for their validity on the relative homogeneity of individual multiples as well as the homogeneity of all instances of multiple invention.'[12] His analysis of the alleged multiple invention of steam turbines and Pelton water wheels suggests 'that both types of identity are illusory'.[13] Constant noted that the turbines of DeLaval and Parsons were different machines designed to respond to separate problems; DeLaval was trying to run a cream separator requiring high r.p.m., while Parsons was trying to generate electricity and required high efficiency. The Curtis and Rateau turbines were also designed to produce electricity, but varied in basic design and number of parts; their variability from the earlier designs was structured in part by attempts to avoid patent laws on existing designs – hardly independent multiple inventions. Also, a major impetus to the evolution of the designs of Parsons, Curtis and Rateau were the requirements of central power stations and electrical power distribution and utilization systems. 'Such products of co-evolutionary meanderings do not seem to be inevitable "multiples" bound to occur someday. Indeed, neither "multiple invention", nor "superorganic cultural elements" nor "social determinism" are conceptually adequate to explain either the origins or the developmental history of steam turbines.'[14]

Constant's conclusion may be applied with equal force to a study of the independent discovery of the asymmetric carbon atom conducted by Hannah Gay.[15] Gay argues convincingly that the discovery *was* made independently by both J. H. van't Hoff and J. A. LeBel. In spite of the fact that they had worked together in Wurtz's laboratory prior to their publications, they had never exchanged views on this matter.

However, as Gay notes, 'it is surely not fortuitous that the two theories appeared at roughly the same time'.[16] After all, 'when many individuals become involved with a particular problem it would be strange for only one of them to arrive at a solution'.[17] The cause of this according to Gay is not some mystical power of cultural maturation like that espoused by Kroeber; in an argument reminiscent of Sapir's criticism of Kroeber, Gay suggests that the outcome is dictated by the structure of nature and persistent anomalies in the research programmes.

Even granting the equivalence of the work of van't Hoff and LeBel, it is interesting to note that as a result of an 1869 paper the priority of E. Palerna, a University of Palermo chemist, has been advocated recently in the pages of the *Journal of Chemical Education*.[18] His priority has generally been overlooked.

Work by Yehuda Elkana on the *Discovery of the Conservation of Energy* has raised related questions.[19] This law with its five independent claimants is frequently cited as the archetypal case of simultaneous discovery. However, Elkana presents strong grounds for concluding that only Helmholtz's contribution should be treated seriously and that the work of others was reconstructed as equivalent from hindsight.

David A. Hounshell's treatment of the simultaneous invention of the telephone by Bell and Gray addresses some different issues in multiple invention.[20] He argues that Bell's prominence in the invention derived from his perception of the potential of a device that could send voice signals. As a teacher to the deaf Bell realized the commercial potential of his new device. On the other hand, though as a professional inventor he understood the mechanism of speech transmission, Elisha Gray was preoccupied with telegraphic equipment and the telegraph business; for him, the new invention was seen as a possible substitute for the telegraphic key. He did not attribute the same commercial potential to his invention and consequently is attributed less prominence in its development.

We see then that multiple discoveries present a very complex mix of attributions made by researchers themselves, their colleagues and their contemporaries, as well as by their successors. All these elements must be parsed out before we can determine the significance of multiple discoveries and inventions for our understanding of technological and scientific innovation.

Latent function of priority disputes

The present theory provides grounds for re-evaluating an important 'latent function' of priority disputes. Specifically, this model recommends that, aside from the normative considerations described by Merton, there are cognitive grounds for these disputes. For example, the dispute involving O. C. Marsh and E. D. Cope, the famous American palaeontologists who described series of new fossil finds in the western badlands, is well known. Though they competed for field hands, waylaid one another's shipments by bribing station masters and local naturalists, and abused one another with mutual charges of intellectual theft, pushing their cases ultimately to the pages of the *New York Herald,* they nonetheless were developing competing classification systems. Consequently, the appropriation of dinosaur specimens had a competing theoretical value, for it could be used to corroborate either the one, or the other's taxonomy. Consequently, the dispute appears to have been grounded at least in part in their competing theories.[21]

Semmelweis' dispute with Scanzoni is interesting for related reasons. Semmelweis, having attributed in 1848 the incidence of childbed fever to a form of 'blood poisoning' from decomposing cadavers, met considerable resistance to his practice of 'prophylaxis', that is, sterilizing the hands before operations by washing them in lime water. Semmelweis tried with little success to propagate his ideas over an eleven-year period following his discovery.[22] However, according to Semmelweis, certain of his critics secretly adopted his methods and experienced success in preventing infections, though they claimed to have arrived at these practices independently. For our purposes, the dispute concerns not who discovered the technique of prophylaxis first, but the sensibility of prophylaxis in the context of the time. Before the rise of the germ theory of infection in later-nineteenth-century medicine, the substantive possibility of Semmelweis' claims were easily disputed. Again, the disputes involving Semmelweis appear to have been rooted in the cognitive structure of science. Only an intensive examination of many cases will be able to differentiate the relative significance of these cognitive as well as normative factors. The current model makes this a highly promising avenue of research.

Rhetoric and method in scientific announcement
A third type of investigation which becomes important in light of the present programme concerns the rhetorical and organizational status of methods in science. For example, in Semmelweis' book on the cause of child-bed fever, a number of possible causes are posited and disputed before the definitive one is disclosed. In fact, most of the 'hypotheses' were never tested at all, but are represented as continuing items in a fictional series of actions whose outcome was the discovery. It is ironic that Carl Hempel cites Semmelweis' case as an instance of

the 'method of hypothesis'.[23] However, it is clear that the method was exploited, not to make the discovery, but as a means of exhibiting its reasonable and valid character to others in and as an announcement. In a parallel vein, G. Nigel Gilbert has outlined some of the subtle ways in which new claims to knowledge are socially constructed in research papers through the transformation of the research record.[24] In other words, from a large though indefinite array of experience, only a selective class of details is relied upon in the production of the announcement. The further investigation of such practices will extend the initial observations outlined earlier regarding the ways in which discoveries have been objectified by our sense making practices, by local traditions and by folk beliefs about how discoveries 'happen'. Though substantively separate, these topics are all animated by the same interest in social construction and objectification.[25]

The group mediation of research

A fourth type of investigation that becomes promising in light of this work is more synthetic in character. A great deal of social research has been directed toward the description of aspects of the scientific community, its productivity, its stratification system, and factors which influence these things. While interest in these topics was spurred by Kuhn's speculations about the existence of paradigmatic camps in science, a number of sociologists have developed more precise ways of describing the network clustering of research groups, the significance of patterns of citation in scientific communication, and factors which influence scientific productivity. The identification of sub-communities is crucial in the context of the attributional model. Sub-communities and their institutions are *mediators* of scientific knowledge. They tend to influence what is considered a problem in the tradition, and hence control the possibility structure of particular innovations. They focus opinions, appraisals and criticisms. They structure the course of socialization for young scientists. They provide regular organized outlets for the announcement and evaluation of innovations by way of conferences and publications. They funnel research funds and structure research priorities. They also influence the nature of the individual scientist's career. Consequently, they tend to mediate all the criteria of discovery discussed in this theory.

The criteria of intelligibility are 'expandables'. While I have treated them for the purposes of this exposition as discrete variables, it could be argued that the degree of consensus regarding plausibility, novelty and validity are matters for continuous measurements, i.e. percentages of a community or a group. Consequently it seems entirely plausible to examine from the standpoint of an organization or community what a discovery looks like, or better still, to specify the characteristics of communities which will influence the attribution of discovery status. For example, the ubiquity of collaboration in the natural sciences, the formation and persistence of research groups and speciality networks as well as the high acceptance rates of journal submissions compared with the social sciences and humanities indicates a far greater consensus about theories and methods in the former compared to the latter communities. This is consequential in that it gives much more vivid relief to innovation; innovative behaviour appears much more discretely defined in natural science communities.[26] Hence, we would expect to find more uniform methods of communication, and smaller variance in the assessment of the importance of particular research, as well as quicker reactions to announcements. These gross organizational features of scientific communities will no doubt amplify, transpose and otherwise influence the assessment process and the labelling of innovation. In other words, the facts of organizational arrangement constitute a set of important intervening variables between the production of innovation and its attribution of discovery status.

Thus the model described here not only presents a theory of discovery which overcomes some of the limitations of previous theories, but indicates the promise of future avenues of related research. I trust that these future pursuits will not only bear out the general thrust of the position advanced here, but will revise and extend the model in light of new investigations. Rather than contradicting it, this prospect will corroborate our understanding of the social basis of discoveries.

Notes

1 The topic of discovery and the concept of nature

1 Norwood Russell Hanson, 'An anatomy of discovery', *Journal of Philosophy*, vol. 64, no. 11, 1967, p. 352.
2 Derek de Solla Price, 'The peculiarity of a scientific civilization', pp. 1–24 in *Science Since Babylon*, enlarged edition, New Haven: Yale University Press, 1975.
3 Cf. Martin Heidegger, *What Is a Thing?*, Chicago: Henry Regnery, 1970, pp. 67ff.
4 Plato, 'The Theatatus', pp. 845–919 in *The Collected Dialogues of Plato*, edited by E. Hamilton and H. Cairns, Princeton: Princeton University Press, 1961, p. 856.
5 René Descartes, *Discourse on Method*, translated by L. J. Lafleur, Indianapolis: Bobbs-Merrill, 1950, pp. 20–6.
6 Cf. William Leiss, *The Domination of Nature*, New York: George Braziller Publishing, 1972, chapter 1.
7 Hans Reichenbach, *Experience and Prediction*, Chicago: University of Chicago Press, 1938, p. 6.
8 Arthur Koestler, *The Act of Creation*, London: Hutchinson, 1964, p. 142.
9 Michael Polanyi, *Personal Knowledge*, Chicago: University of Chicago Press, 1958, pp. 110ff.
10 Cf. Koestler, *The Act of Creation*, pp. 169–70.
11 Cf. Henry Bett, *Nicholas of Cusa*, London: Methuen, 1932, pp. 176ff; also see Karl Jaspers, *Anselm and Nicholas of Cusa*, edited by Hannah Arendt, translated by Ralph Manheim, New York: Harcourt, Brace and World, 1966, pp. 105ff.
12 Karl Popper, *The Logic of Scientific Discovery*, New York: Basic Books, 1959, pp. 31–2.
13 Richard Braithwaite, *Scientific Explanation*, Cambridge: Cambridge University Press, 1953, pp. 20–1.
14 Carl R. Kordig, *The Justification of Scientific Change*, Dordrecht: Reidel, 1971.
15 R. G. A. Dolby, 'The sociology of knowledge in natural science', pp. 309–20 in *Sociology of Science*, edited by Barry Barnes, Harmondsworth: Penguin Books, 1972, p. 313.
16 David L. Hull, *Darwin and His Critics*, Cambridge, Mass.: Harvard University Press, 1973, p. 455. Paraphrased.
17 Gerald Holton, 'Einstein, Michelson and the "crucial" experiment', pp. 261–352 in *Thematic Origins of Scientific Thought*, Cambridge, Mass.: Harvard University Press, 1973, p. 278.
18 Hans Reichenbach quoted in *ibid.*, p. 278.
19 *Ibid.*, p. 279.
20 Paul K. Feyerabend, *Against Method*, London: New Left Books, 1975, p. 141.
21 *Ibid.*, p. 154.

22 *Ibid.*, p. 153.
23 *Ibid.*, pp. 155–6.
24 *Ibid.*, p. 165.
25 *Ibid.*, p. 166.
26 Ludovico Geymonat, *Galileo Galilei*, translated by Stillman Drake, New York: McGraw Hill, 1965 (first Italian edition 1957), p. 38. Cited in *ibid.*, p. 142.
27 The term 'theory-loaded' appears to have been coined by Norwood Russell Hanson in *Patterns of Discovery*, London: Cambridge University Press, 1958. However, later philosophers adopted the term 'theory-laden'. Cf. Kordig, *Justification of Scientific Change*.
28 Thomas S. Kuhn, *The Structure of Scientific Revolutions*, second edition, Chicago: University of Chicago Press, 1970, p. 9.
29 Gerald Holton, 'The thematic imagination in science', pp. 47–68 in *Thematic Origins*.
30 Wilhelm Wien, cited in Holton, *Thematic Origins*, p. 269.
31 Albert Einstein, cited in Holton, *Thematic Origins*, p. 237.
32 Feyerabend, *Against Method*, chapter 1. Though it is often common within orthodox circles of the philosophy of science to lump Feyerabend, Kuhn and Hanson together as a sort of lunatic fringe, one should be aware of some basic differences in emphasis. Specifically, Feyerabend has eschewed the notion of the 'theory-loaded' character of observations in favour of an extended investigation of incommensurability between rival research traditions. While the advocates of the theory-loaded observation school have difficulty in accounting for scientific change, Feyerabend points to the role of internal contradictions and subjective wishes as sources of change in a content of incommensurable theories. See *ibid.*, chapter 17.
33 Holton, *Thematic Origins*, pp. 12ff.

2 *Psychological accounts of discovery*

1 Norwood Russell Hanson, 'Notes toward a logic of discovery', pp. 42–65 in *Perspectives on Peirce*, edited by R. J. Bernstein, New Haven: Yale University Press, 1965, p. 48. Also see Hanson's *Patterns of Discovery*, Cambridge: Cambridge University Press, 1958, p. 17.
2 This three part classification is offered by Hanson himself in 'Notes toward a logic of discovery', pp. 46–9. 'Is there a logic of discovery' appeared in *Current Issues in the Philosophy of Science*, edited by H. Feigl and G. Maxwell, New York: Holt, Rinehart and Winston; 'Retroductive inference' appeared in *Philosophy of Science: The Delaware Seminar*, vol. 1 (1961–2), edited by Bernard Baumrim, New York: Wiley, 1963.
3 Hanson, *Patterns of Discovery*, pp. 5ff.
4 *Ibid.*, p. 30.
5 Hanson, 'Is there a logic of discovery?'.
6 Hanson, 'Retroductive inference'.
7 Hanson, 'Notes toward a logic of discovery', p. 50.
8 Norwood Russell Hanson, 'An anatomy of discovery', *Journal of Philosophy*, vol. 64, no. 11, 1967, pp. 321–52. It should be noted that Hanson's orientation to Wittgenstein was not a later development. His books on the structure of discovery and on the discovery of the positron are preoccupied with the conceptual bases of both

traditional theories and innovative thinking, especially in physics. The same cannot be said of his papers on the logic of discovery. Consequently, I share much of the same orientation to the conceptual basis of the 'empirical world', though we differ considerably on what a study of the conceptual basis of discovery should look like. Though Hanson's major work is entitled *Patterns of Discovery*, this is misleading. The book deals with the idea structures which ground observations and experiments, which direct inferences and suggest alternative accounts, and which warrant the comparability of tests, observations and theories within any particular tradition. It is a powerful treatment of the coherence and integrity of such diverse matters within any tradition, but is less forceful in explaining change between traditions. His comparative treatment of classical particle physics versus elementary particle physics gives relief to the conventional character of scientific thought and raises questions about the complementarity of the two traditions. In this respect, Hanson develops a much more 'internalistic' approach than my own. For my part, the initial problem is to elucidate the conceptual basis of our common notion of 'discovering' something, and then to describe how this applies in science. Hanson undertook the more ambitious route of describing the idea basis of scientific concepts like the positron, inertia, uncertainty, etc. Consequently, Hanson's work captures scientists as natural philosophers struggling with concepts that organize nature. My work relates to scientists as ordinary members of society making common sense attributions in a technical field. These efforts are complementary. Hanson's difficulties arise when he attempts to spell out a theory of innovative behaviour.

It should be pointed out parenthetically that in this text all references to scientists as 'members of society' or 'collectivity members' are somewhat elliptical. Membership means primarily a solidarity with a collectivity or social group. However, in my usage what is important is that this solidarity consists of *the knowledge* of the common sense and the taken for granted organization of settings which is recognized in common among the various 'natives' to a setting. Consequently, talk about scientists as members of society is not meant to indicate an actor with some general familiarity with the facts of life, but a person with an expertise in the organizations and social structures of his everyday life. The scientist as a collectivity member has a common sense knowledge of theories, journals, universities, research organizations, funding agencies and the like. While his *substantive* knowledge of the world will differ with the layman's, this is beside the point. The concept of member points to the *quality* of his knowledge, specifically that by virtue of his training and experience, it is typically intuitive, spontaneous and taken for granted in the daily conduct of scientific life.

Though this usage of 'member' is clearly technical, authors in interpretive sociology rarely make this clear. The single most important source for the discussion of these issues is Harold Garfinkel, *Studies in Ethnomethodology*, Englewood Cliffs: Prentice Hall, 1967, chapters 2 and 3.

9 Richard J. Blackwell, *Discovery in the Physical Sciences*, Notre Dame: Notre Dame University Press, 1969.

10 *Ibid.*, p. 18.

11 *Ibid.*, pp. 18–24.

12 *Ibid.*, p. 78.

13 *Ibid.*

14 *Ibid.*, p. 87–8.

15 *Ibid.*, p. 101.

16 *Ibid.*, p. 103.

17 *Ibid.*, p. 152.

18 *Ibid.*, p. 149. By structures, Blackwell means 'the constructs of thought' or ideas. The five point process is synthesized from longer discussions by Blackwell, *ibid.*, pp. 129–48.

19 The three point process is synthesized from longer discussions by Blackwell, *ibid.*, pp. 166–82.

20 *Ibid.*, p. 17.

21 *Ibid.*, p. vii.

22 *Ibid.*, p. 19.

23 *Ibid.*, p. 29.

24 *Ibid.*, p. 103.

25 *Ibid.*

26 Hanson, 'An anatomy of discovery', p. 350.

27 Norwood Russell Hanson, 'The logic of discovery', *Journal of Philosophy*, vol. 55, no. 25, 1958, pp. 1073ff.

28 *Ibid.*, p. 1089.

29 *Ibid.*, p. 1088.

30 Thomas S. Kuhn, 'The function of dogma in scientific research', pp. 59–89 in *Scientific Change*, edited by A. C. Crombie, London: Heinemann, 1963.

31 Thomas S. Kuhn, *The Structure of Scientific Revolutions*, second edition, Chicago: University of Chicago Press, 1970, p. 52.

32 Cf. J. S. Bruner and Leo J. Postman, 'On the perception of incongruity', *Journal of Personality*, vol. 18, 1949, pp. 206–23.

33 Thomas S. Kuhn, 'Historical structures of scientific discoveries', *Science*, vol. 136, 1962, p. 763.

34 *Ibid.*

35 'Scientists repeatedly revise their expectations, usually their instrumental standards, and sometimes their theories as well.' *Ibid.*, p. 763.

36 *Ibid.*, p. 761.

37 Kuhn, *Structure of Scientific Revolutions*, p. 53.

38 Quoted from Herschel's diary and cited in Kuhn, 'Historical structures of scientific discoveries', p. 762.

39 Cf. a similar point in Bernard Barber, 'The social process of invention and discovery', chapter 9 in *Science and the Social Order*, New York: Macmillan, 1952, p. 268.

40 Kuhn, *Structure of Scientific Revolutions*, p. 65.

41 Kuhn, 'Historical structures of scientific discoveries', p. 760.

42 *Ibid.*, p. 763: 'There is no single moment or day which the historian, however complete his data, can identify as the point at which the discovery was made.'

43 *Ibid.*, p. 761. The researcher's *own* view of his activities, i.e. his or her attribution of importance to the innovation is all important. This has been discussed by several writers. For example, see Edward Constant II: 'paradigmatic commitment shapes not only what an inventor invents, but how he recognizes it'; p. 203 in 'On the diversity and co-evolution of technological multiples', *Social Studies of Science*, vol. 8, 1978, pp. 183–210.

44 Kuhn, 'Historical structures of scientific discoveries', p. 761.

45 *Ibid.*

46 Indeed, mention of Bayen and Hales is omitted entirely from the first edition of Kuhn, *Structure of Scientific Revolutions*, though they appear significantly in his article of the same year, 'Historical structures of scientific discoveries'.

47 Wesley C. Salmon, *Logic*, Englewood Cliffs: Prentice-Hall, 1963, p. 28. The form of the argument is

    if p, then q

    q,

    therefore p.

This is called the fallacy of affirming the consequent. Substituting the appropriate statements we get the following:

    if anomalies appear in the course of scientific research, then discoveries will be made

    discoveries have been made

    therefore anomalies produced the scientific discoveries.

The concluding statement is erroneously deduced from the premises.

48   if p, then q

    p is the case

    therefore q is the case

This is a valid deduction called 'affirming the antecedent', cf. *ibid.*, p. 2.

49 Arthur Koestler, *The Act of Creation*, London: Hutchinson, 1964, pp. 32–3.

50 *Ibid.*, pp. 64–86.

51 *Ibid.*, p. 123.

52 *Ibid.*, pp. 131–44.

53 *Ibid.*, pp. 124–50.

54 This quotation is from Henri Poincaré, *Science and Method*, translated by Francis Maitland, New York: Scribner, 1914, and is cited throughout the literature on creativity, cf. *ibid.*, pp. 115–16.

55 *Ibid.*, p. 154.

56 Einstein cited in *ibid.*, pp. 146, 158.

57 Michael Polanyi, *The Tacit Dimension*, Garden City: Doubleday, 1966.

58 Michael Polanyi, *Personal Knowledge*, Chicago: University of Chicago Press, 1958.

59 Indeed, it is this 'vacuity' which allows us to switch freely from topic to topic. From this point of view, we may switch in and out of gestalt sets automatically. I go for my ticket, find it, stay in line to board the bus, a solution to an earlier problem pops into mind, the driver tells me to go 'right to the back' and I reflect a moment on my discovery, then pick up on the conversation with a friend.

These *in situ* switches in consciousness are not of the sort implied by the textbook images of perceptual gestalt. What is switched here is the topic of focal attention. It is not as though the idea has come 'out of the blue'. As Poincaré notes, he had been consciously immersed with the problem of Fuchsian-type functions before he arrived at his discovery. The fact that he arrived at the solution when he did indicates that he still enjoyed a level of involvement or a 'preoccupation' with the matter, in spite of the fact that this was not purely and fully conscious. This kind of subsidiary involvement is obscured by the textbook examples of figure–ground dependence.

Also, Poincaré's account suggests that gestalt switches do not mutually exclude one another structurally. In the textbook example, the perception of the goblet excludes structurally the presence of the faces and vice versa (i.e. the figure of one 'grounds' the other image). However, this is not the case in Koestler's account of Poincaré and Archimedes. A theme may be *inconsistent* with a course of activity; the discovery might disrupt the bath, or the conversation might disrupt further dwelling on the discovery. This seems to be the case because we can only 'pay attention' to one thing at a time – even though over a few seconds we may be struck by several thematically independent matters, one after another in quick succession.

However, the presence of the one matter does not structurally obliterate the other. Indeed, the gestalt switch is an *integrative* event which makes present past problems that have been 'kept alive' in the subconscious mind, and in the case of discovery, have been the subject of non-conscious 'reflection'. This is altogether different from the textbook examples of gestalt which give little play to memory, and consequently where the integrative significance of the shift in consciousness is obscured by an 'optical illusion'.

60 Alfred Schutz, 'On Multiple Realities', pp. 207–59 in *The Collected Papers*, vol. 1: *The Problem of Social Reality*, The Hague: Martinus Nijhoff, 1962.

61 Erving Goffman, *Frame Analysis*, New York: Colophon Books, 1974.

62 Consequently, though the activity which the shift shifts may in the immediate moment be 'off-topic', its inconsistency with the matter at hand is merely contingent. Unlike that variety of gestalts which are optical illusions, the presence of one idea in an *in situ* gestalt does not alter the substance or form of what it displaces. In the case of Archimedes, the discovery terminates the *episode* of bathing at that point, but bathing as such is not recast, though Archimedes' attention is certainly elsewhere.

63 Koestler, *The Act of Creation*, pp. 210–11.

64 *Ibid.*, p. 112.

65 Ignaz Semmelweis, *Aetiology, Concept and Prophylaxis*, 1861, translated and reprinted in part in W. J. Sinclair, *Semmelweis: His Life and His Doctrine*, Manchester: Manchester University Press, 1909.

66 Cf. Hanson, *Patterns of Discovery*, pp. 203–4.

67 Poincaré, 'Mathematical Discoveries', in *Science and Method*, pp. 51ff.

68 Cf. Gerald Holton, 'Poincaré and Relativity', pp. 185–95 in *Thematic Origins of Scientific Thought*, Cambridge, Mass.: Harvard University Press, 1973.

69 Cf. Leslie White, *The Science of Culture*, second edition, New York: Farrar, Straus and Giroux, 1969, pp. 206–7.

3 *A synthetic assessment of the psychological accounts*

1 See for example Herbert A. Simon, *Models of Discovery and Other Topics in the Methods of Science*, Dordrecht: Reidel, 1977.

2 Cf. Norwood Russell Hanson, *Patterns of Discovery*, Cambridge: Cambridge University Press, 1958, pp. 203–4.

3 Cf. Gerald Holton, 'Poincaré and Relativity', pp. 185–95 in *Thematic Origins of Scientific Thought*, Cambridge, Mass.: Harvard University Press, 1973.

4 Cf. Max Wertheimer, 'On truth', *Social Research*, vol. 1, 1934, pp. 135–46; 'Gestalt theory', *Social Research*, vol. 11, 1944, pp. 78–99; *Productive Thinking*, New York: Harper, 1945.

5 Cf. Thomas S. Kuhn, 'Historical structures of scientific discoveries', *Science*, vol. 136, 1962, p. 762. Also see L. W. Taylor, *Physics, the Pioneer Science*, Boston: Houghton Mifflin, 1941, pp. 790ff.

6 Kuhn, 'Historical structures of scientific discoveries'.

7 Compare Thomas S. Kuhn, *The Structure of Scientific Revolutions*, second edition, Chicago: University of Chicago Press, 1970, p. 35, and Kuhn, 'The essential tension: tradition and innovation in scientific research', pp. 341–54 in *Scientific Creativity, Its Recognition and Development*, edited by Calvin W. Taylor and Frank Barron, New York: Kreiger, 1963, p. 342.

8 Galileo Galilei, 'The Starry Messenger' and 'Letters on Sunspots' in *The Discoveries and Opinions of Galileo,* translated and edited by Stillman Drake, New York: Anchor Books, 1957.

9 Cf. S. W. Williston, 'The first discovery of dinosaurs in the west', pp. 124–31 in *Dinosaurs,* edited by W. D. Matthews, New York: American Museum of Natural History, 1915.

10 Sigmund Freud, *Leonardo da Vinci,* translated by Alan Tyson, New York: Norton, 1964; Lewis S. Feuer, *Einstein and the Generations of Science,* New York: Basic Books, 1974.

11 Hans Reichenbach, *Experience and Prediction,* Chicago: University of Chicago Press, 1938, pp. 6–7.

12 This was suggested to me by Professor Fred Wilson, Department of Philosophy, University of Toronto.

13 Karl Mannheim, *Ideology and Utopia,* New York: Harcourt, Brace and World, 1936, pp. 282–4.

14 Karl Popper, *The Poverty of Historicism,* second edition, London: Routledge and Kegan Paul, 1960.

15 *Ibid.,* p. 158.

16 Wertheimer, *Productive Thinking,* p. 212.

17 Richard J. Blackwell, *Discovery in the Physical Sciences,* Notre Dame: Notre Dame University Press, 1969, p. 103.

18 Kuhn, *Structure of Scientific Revolutions,* p. 53; also see Uno Bocklund, 'A lost letter from Scheele to Lavoisier', *Lychnos,* vol. 39, 1957–8, pp. 39–62. See also J. R. Partington, 'The discovery of oxygen', *Journal of Chemical Education,* vol. 39, no. 3, 1962, pp. 123–5. As Partington points out, Scheele's work, though it may have been delayed in publication due to circumstances beyond his control, was nonetheless reported in some detail in a memoir written by T. Bergman in 1775.

19 These cases are discussed at length in later chapters.

4 The emergence of a social model of discovery

1 Cf. Cynthia Eagle Russett, *Darwin in America,* San Francisco: Freeman, 1976, pp. 83–124; also see Richard Hofstadter, *Social Darwinism in American Thought,* Philadelphia: University of Pennsylvania Press, 1944.

2 This was not strictly speaking a 'debate', for Galton had already died before most of these others had registered their objections, and had drawn on the record of multiple discoveries. Also, it is unclear that there would have been a debate had Galton been alive – for he had himself observed the frequent occurrence of multiple discoveries. This however is never dealt with in the anthropological literature.

3 William F. Ogburn and Dorothy Thomas, 'Are inventions inevitable? A note on social evolution', *Political Science Quarterly,* vol. 37, no. 1, 1922, pp. 83–98; reprinted in William F. Ogburn, *Social Change* New York: Heubsch, 1922, pp. 80–102.

4 Ogburn, *Social Change,* p. 343.

5 Leslie White, *The Science of Culture,* second edition, New York: Farrar, Straus and Giroux, 1969, p. 207, footnote.

6 *Ibid.,* pp. 206–7.

7 *Ibid.,* p. 214.

8 *Ibid.,* p. 214.

9 Francis Galton, *Finger Prints,* London: Macmillan, 1892.

10 Francis Galton, *Hereditary Genius,* London: Macmillan, 1869; reprinted Gloucester, Mass.: Meridian Books, 1962. Also see 'Francis Galton', in William H. Kruskal and Judith M. Tanun (eds.), *International Encyclopedia of Statistics,* vol. 1, New York: Macmillan, 1968, pp. 359–64. Also see entry for 'Karl Pearson', vol. 2, pp. 691–8.

11 Galton, *Hereditary Genius,* p. 331.

12 Charles Horton Cooley, 'Genius, fame and the comparison of races', *Annals of the American Academy of Political and Social Science,* vol. 9, 1897, pp. 1–42.

13 A. L. Kroeber, 'The superorganic', *American Anthropologist,* vol. 19, no. 2, 1917, p. 167.

14 We noted in chapter 3 that mentalistic theories tended to be post hoc. This is also the case with Galton. By starting with a sample of already successful families, he cannot control for which successes are occurring genetically, and which are occurring otherwise. So too, because he does not begin with a sample at large, he cannot examine cases where hereditary genius has no marks of distinction.

15 Kroeber, 'The superorganic', p. 200.

16 *Ibid.,* p. 197.

17 *Ibid.,* p. 198.

18 *Ibid.,* p. 199.

19 *Ibid.,* p. 202.

20 White, *The Science of Culture,* p. 292.

21 Cf. H. M. Blalock Jr, *Theory Construction,* Englewood Cliffs: Prentice-Hall, 1969, p. 91: 'Expressing a change in x as a function of time is an admission of one's ignorance as to the causal dynamics.'

22 White, *The Science of Culture,* p. 226.

23 *Ibid.,* p. 230.

24 *Ibid.,* p. 229.

25 Kroeber, 'The superorganic', p. 200.

26 Robert K. Merton, *The Sociology of Science, Theoretical and Empirical Investigations,* edited and with introductory remarks by Norman W. Storer, Chicago: University of Chicago Press, 1973. We were especially interested in chapters 13–21; this quote is from chapter 17, 'Multiple discoveries as strategic research site', p. 371.

27 Norman W. Storer, prefatory note to Section 4, 'The reward structure of science', in Merton, *The Sociology of Science,* p. 281. It is of some significance that although the five chapters in this section are concerned with scientific discovery, the section heading emphasizes *reward:* 'The reward system of science'. This suggestion is only loosely borne out by the articles themselves.

28 Merton, *The Sociology of Science,* 'Singletons and multiples in science', chapter 16, p. 356.

29 *Ibid.,* p. 357.

30 *Ibid.,* p. 358.

31 *Ibid.,* p. 359.

32 *Ibid.,* p. 359.

33 *Ibid.,* pp. 361–4; there is some reason to doubt the significance of sealed notes kept on deposit with academies of science as we shall see in the next chapter. It should also be pointed out that *fear* of duplication is quite a different matter from actual duplication. Warren Hagstrom found that while a great many scientists fear being anticipated, this had actually happened to less than 5% of those sampled. Cf. Warren O. Hagstrom, *The Scientific Community,* New York: Basic Books, 1965, p. 70.

34 Merton, *The Sociology of Science,* p. 366.

35 *Ibid.,* p. 352.
36 *Ibid.,* 'Priorities in scientific discovery', chapter 14.
37 *Ibid.,* p. 293.
38 *Ibid.*
39 *Ibid.,* p. 300.
40 Cf. William K. Stuckey, 'The prize', *The Saturday Review,* vol. 55, no. 36, 1972, pp. 33–9; also see Harriet Zuckerman's definitive work on Nobel laureates, *The Scientific Elite,* New York: Free Press, 1977.
41 Merton, *The Sociology of Science,* p. 303.
42 *Ibid.,* p. 305.
43 Jonathan Cole and Stephen Cole, *Social Stratification in Science,* Chicago: University of Chicago Press, 1973; Jerry Gaston, *Originality and Competition in Science,* Chicago: University of Chicago Press, 1973; Harriet Zuckerman, 'Nobel laureates in science', *American Sociological Review,* vol. 32, 1967, pp. 391–403; Zuckerman, 'The sociology of the Nobel prize', *Scientific American,* vol. 217, no. 5, 1967, pp. 25–33; Zuckerman, 'Patterns of name-ordering among authors of scientific papers', *American Journal of Sociology,* vol. 74, no. 4, 1968, pp. 276–91.
44 Merton, *The Sociology of Science,* pp. 371–82.
45 *Ibid.,* p. 289.
46 On the other hand, the idea that discoveries are in principle multiples suggests that their social definition is an irrelevant consideration. The fact that they repeatedly occur simultaneously makes such appearances seem naturalistic and inevitable, and hence belittles any consideration of their social construction. However, the behaviour of scientists, especially during priority disputes, indicates that discoveries are not held to be multiples.
47 Merton, *The Sociology of Science,* 'The ambivalence of scientists', ch. 18, p. 400. In a later programmatic paper, Merton does identify an ambiguity in the sense of 'recognition' similar to the one discussed here. However, this does not seem to have been influential in developing a theory of the attributional process. See ' "Recognition" and "excellence": instructive ambiguities', ch. 19 in *ibid.,* pp. 419–38. This paper originally appeared in 1960.
48 Compare this to the record left by the Phoenicians atop Beauvoir Mountain near Sherbrooke, Quebec in the fifth century B.C. Apparently they had crossed the Atlantic via the Canary Islands two thousand years before Columbus, had entered the St Lawrence River and found a land theretofore undiscovered by the ancients. The archaeological data show that records of their discoveries were left, not only near Sherbrooke, but also in the Canary Islands and near the Mojave Desert in Utah.

   The discoveries were recorded by splitting a large rock and cutting an inscription into the exposed faces, and finally fitting together the two fractured pieces, binding them and leaving them in a geographically prominent location. The stone found near Sherbrooke bears the following inscription: 'Thus far our expedition travelled in the service of Lord Hiran, to conquer land. This is the record of Hanta, who attained the great river, and these words cut on stone.' These records were left to announce the accomplishment of the Phoenicians. They were left to have the achievement *recognized,* not in the sense of having the argonauts pensioned and titled, but in the sense of having their presence *acknowledged.* Unlike Columbus, the Phoenician record did not appear in a context where recognition of the announcement could be mobilized institutionally. When the stone records were found some two hundred years ago by an early Quebec farmer, they were apparently seen as striking curiosities, for the farmer displayed the stones as monuments in front of his house. They came to the attention of the archaeological community only at about the turn of the century; and only in the last two years has the script been identified, dated and translated. In the case of the Phoenicians, our recognition of their accomplishment consisted in locating the stone, assigning it a historical place, and reading it as a factual record of a local discovery. In this sense, social recognition here refers to the recognition that something was a discovery – as opposed to its ceremonial approval in scientific and political institutions. The latter is clearly secondary in that it is tied exclusively to the former, more basic sense of recognition.

Cf. C.B.C. Ideas, 'The Sherbrooke Stones', materials distributed for a March 6 C.B.C. Radio broadcast, Toronto, Canadian Broadcasting Corporation Press, 1977. Also see Barry Fell, *America B. C.*, New York: Quadrangle Books, 1976. Though there is some controversy over the authenticity of these stones, the point I am making should be nonetheless quite obvious.

49 Dorothy Thomas' reservations about the multiples will be explored in more detail in chapter 8.

5 *Discovery as meaningful action*

1 David Bloor, *Knowledge and Social Imagery*, London: Routledge and Kegan Paul, 1976, p. 6.

2 Several of the representatives of this broad school are Barry Barnes, *Scientific Knowledge and Sociological Theory*, London: Routledge and Kegan Paul, 1974; B. Barnes and S. Shapin (eds.), *Natural Order: Historical Studies of Scientific Culture*, Beverly Hills and London: Sage, 1979; Bloor, *Knowledge and Social Imagery*; H. M. Collins, 'The seven sexes: a study in the sociology of a phenomenon, or the replication of experiment in physics', *Sociology*, vol. 9, 1975, pp. 205–24; David O. Edge and Michael J. Mulkay, *Astronomy Transformed: The Emergence of Radio Astronomy in Britain*, New York: Wiley Interscience, 1976; G. N. Gilbert, 'The transformation of research findings into scientific knowledge', *Social Studies of Science*, vol. 6, 1976, pp. 281–306; John Law and David French, 'Normative and interpretive sociologies of science', *Sociological Review*, vol. 22, 1974, pp. 581–95; Michael J. Mulkay, *Science and the Sociology of Knowledge*, London: Allen and Unwin, 1979; Bruno Latour and Steve Woolgar, *Laboratory Life, The Social Construction of Scientific Facts*, Beverly Hills and London: Sage, 1979; B. Wynne, 'C. G. Barkla and the J. phenomenon: a case study of the treatment of deviance in physics', *Social Studies of Science*, vol. 6, 1976, pp. 307–47.

3 Robert W. Mackay, 'Conceptions of children and models of socialization', pp. 180–93 in *Ethnomethodology*, edited by Roy Turner, Harmondsworth: Penguin, 1974; also see A. Cicourel, et al., *Language Use and School Performance*, New York: Academic Press, 1974.

4 Cf. Robert Mackay, 'The ethnography of the classroom', unpublished Ph.D. dissertation, Department of Sociology, University of California at Santa Barbara, 1974.

5 This example is given by L. Wittgenstein, *The Blue and Brown Books*, Oxford: Basil Blackwell, 1958, p. 23.

6 Peter Winch, *The Idea of a Social Science and Its Relation to Philosophy*, London: Routledge and Kegan Paul, 1958.

7 G. E. M. Anscombe, 'Causality and determination', pp. 63–81 in *Causation and Conditionals*, edited by E. Sosa, Oxford: Oxford University Press, 1975, p. 68.

8 *Ibid.*, pp. 68–9.

9 Winch, *The Idea of a Social Science*, p. 15.

10 *Ibid.*, p. 51.

11 *Ibid.*, p. 32.

12 *Ibid.*, p. 59.

13 Cf. Zhores Medvedev, *The Rise and Fall of T. D. Lysenko*, Garden City: Doubleday, 1971; also see D. Lecourt, *Proletarian Science*, translated by Ben Brewster, London: New Left Books, 1977.

14 Nicolaas A. Rupke, '*Bathybius haeckelii* and the psychology of scientific discovery', *Studies in the History and Philosophy of Science*, vol. 7, no. 1, 1976, pp. 53–62.

15 See Jean Rostand, *Error and Deception in Science*, translated by A. R. Pomerans, London: Hutchinson, 1960. An excellent review of the historical materials was provided recently by Irving M. Klotz, 'The N-ray affair', *Scientific American*, vol. 242, no. 5, 1980, pp. 168–75.

16 Charles Darwin, 'An historical sketch', originally appeared as a preface to the third and later editions of *The Origin of Species* in 1861. Reproduced in P. Appleman (ed.), *Darwin, A Norton Critical Edition*, New York: Norton, 1970, pp. 28–35.

17 See the discussion of Robert Chamber's *Vestiges of the Natural History of Creation* (which beginning in 1844 went through eleven editions amounting to about 24,000 copies) in Milton Millhauser, *Just Before Darwin*, Middletown: Wesleyan University Press, 1959.

18 Lewis S. Feuer, *Einstein and the Generations of Science*, New York: Basic Books, 1974, pp. 67–70. 'When Bertrand Russell published an essay in the British philosophical periodical organ, *Mind*, entitled "Is Position in Time and Space Absolute or Relative?", he acknowledged Moore: "The logical opinions which follow are in the main due to Mr G. E. Moore, to whom I also owe my first perception of the difficulties in the relational theory of space and time" ' (p. 69). 'Russell wrote with assurance that the arguments against absolute time were based on an "antiquated logic" capable of "an easy and simple refutation"; he defined "event" and "simultaneity" in an absolutist sense' (p. 68).

'Whitehead treated time from the absolutist point of view, referring to the young Russell's article for support. "It seems most unlikely" wrote Whitehead, "that existing physicists would, in general, gain any advantage from deserting familiar habits of thought". Five years later in 1910, in an article for the *Encyclopedia Britannica*, Whitehead wrote that as far as the relational and absolute theories of space were concerned, "no decisive argument for either view has been elaborated" ' (p. 69).

Feuer argues further that the absolutism of Moore insulated the Cantabrigians from the social philosophy of Marx, while the relativism of Marxism and Machian physics which enjoyed such great currency among the Zurich–Berne intellectuals favourably disposed the latter toward the *a priori* validity of relativity theory. Einstein, as a member of the Zurich group, was consequently sympathetic to a relativistic outlook even before his 1905 papers.

It should be pointed out that Professor Feuer's interpretation has been the subject of some discussion and clarification. See Joseph Agassi, 'In search of the zeitgeist', *Philosophy of the Social Sciences*, vol. 5, 1975, pp. 339–42. Also in the same journal see Professor Feuer's discussion in 'Method in the sociology of science: rejoinder to Professor Agassi', vol. 6, 1976; pp. 249–53; 'Historical method in the sociology of science: the pitfalls of a polemicist', vol. 7, 1977, pp. 255–61.

Also it was suggested to the author by a reviewer of this manuscript that Russell had developed a strong anti-relativist position *before* Moore's work on ethics. However, as Feuer points out, several Cantabrigians observed the co-emergence of these views. J. M. Keynes wrote in the following terms: 'We combined a dogmatic treatment as to the nature of experience with a method of handling it which was extravagantly scholastic. Russell's *Principles of Mathematics* came out in the same year (1903) as *Principia Ethica;* and the former, in spirit, furnished a method of handling the material provided by the latter' (quoted in Feuer, *Einstein,* p. 68).

In other words, Moore's ethics and Russell's method were seen to be interrelated. Yet it ought to be pointed out that the Cantabrigians were very quick to support the special theory of relativity, especially in 1919 after Eddington's observations during the solar eclipse.

It has also been conjectured that there was a certain patriotic and nationalistic element which anticipated the experiment; many viewed the impending Eddington expedition as an opportunity for the British to show that 'Newton was right and that German and Jewish chap all wrong' (in the words of my anonymous reviewer). While the evidence for this has yet to be gathered, we do know that as a *result* of the experiment, Einstein's theory became quite orthodox.

19 Arthur Koestler, *The Case of the Midwife Toad,* London: Hutchinson, 1971.

20 Rupke, '*Bathybius haeckelii',* pp. 54–5, 59.

21 Adrian Desmond, *The Hot-Blooded Dinosaurs,* London: Blond and Briggs, 1975.

22 E. T. Bell, *Men of Mathematics,* New York: Simon and Schuster, 1937, p. 376.

23 See the Linnean Society papers, pp. 81–97 in Appleman, *Darwin, A Norton Critical Edition,* especially p. 82.

24 This case is discussed by Hans Selye, *From Dream to Discovery,* New York: McGraw Hill, 1964, pp. 91–2.

25 See Charles Darwin, *Autobiography and Selected Letters,* edited by Francis Darwin (1892), New York: Dover, 1958, pp. 180–2.

26 This episode is recorded in the entry for February 14, 1493 during the first voyage. The journals of the explorations are edited and translated by Cecil Jane, *The Voyages of Christopher Columbus,* London: The Argonaut Press, 1930, pp. 245–7.

27 See Darwin, *The Autobiography,* p. 196.

28 Barnes, *Scientific Knowledge and Sociological Theory,* p. 43.

29 *Ibid.,* p. 67.

30 Bloor, *Knowledge and Social Imagery* p. 2.

31 David Bloor, 'Wittgenstein and Mannheim on the sociology of mathematics', *Studies in the History and Philosophy of Science,* vol. 4, no. 2, 1973, pp. 173–91.

32 Barnes, *Scientific knowledge and Sociological Theory,* 'The culture of the natural sciences', pp. 45–68.

33 *Ibid.,* p. 49.

34 *Ibid.,* p. 70.

35 Anscombe, 'Causality and determination', pp. 75–6.

36 Karl Mannheim, *Ideology and Utopia,* translated by Louis Wirth and Edward A. Shils, New York: Harcourt, Brace and World, 1936, pp. 78–9.

37 Steve Woolgar, 'Writing an intellectual history of scientific development: the use of discovery accounts', *Social Studies of Science,* vol. 6, 1976, pp. 395–422. Also see F. G. Smith and A. Hewish (eds.), *Pulsating Stars,* London: Macmillan, 1968.

38 Edge and Mulkay, *Astronomy Transformed,* p. 229.

39 Law and French, 'Normative and interpretive sociologies of science'.

40 Sir Edmund Whittaker, *A History of the Theories of Aether and Electricity: The Modern Theories 1900–1926,* London: Nelson, 1953. Also see Gerald Holton, *Thematic Origins of Scientific Thought,* Cambridge, Mass.: Harvard University Press, 1973, pp. 165–83.

41 Cf. Harold Garfinkel, *Studies in Ethnomethodology,* Englewood Cliffs: Prentice-Hall, 1967.

6 *The law valid for* Pisum *and the reification of Mendel*

1 Cf. Bernard Barber's famous discussion, 'Resistance by scientists to scientific discovery', *Science,* vol. 134, 1961, pp. 596–602; E. G. Boring, *History, Psychology and Science, Selected Papers,* edited by R. J. Watson and D. T. Campbell, New York: Wiley, 1963, p. 33, p. 177; Loren Eiseley, *Darwin's Century,* New York: Anchor, 1961; Elizabeth Gasking, 'Why was Mendel's work ignored?', *Journal for the History of Ideas,* vol. 20, 1959, pp. 60–84; Garrett Hardin, *Nature and Man's Fate,* New York: Holt, Rinehart and Winston, 1959; Gunther S. Stent, 'Prematurity and uniqueness in scientific discovery', pp. 94–114 in *Paradoxes of Progress,* San Francisco: Freeman, 1978. These writers are only a small sample of those who advocate the story of Mendel's 'long neglect'.
2 Thomas S. Kuhn suggests not only that there are two types of discoveries in science, 'normal' and 'revolutionary', but that a particular achievement might constitute a normal discovery for one group while being a revolutionary discovery for another. See Kuhn, *The Structure of Scientific Revolutions,* 2nd edn, Chicago: University of Chicago Press, 1970, p. 51.
3 See Herbert Wendt, *In Search of Adam,* translated by James Cleugh, New York: Collier Books, 1962, pp. 358–9.
4 Lindley Darden, 'Reasoning in scientific change: Charles Darwin, Hugo de Vries, and the discovery of segregation', *Studies in the History and Philosophy of Science,* vol. 7, no. 2, 1976, pp. 154ff.
5 Hugo de Vries, 'The segregation of hybrids', translation by Evelyn Stern of 'Das Spaltungsgesetz der Bastarde' in Curt Stern and Eva R. Sherwood (eds.), *The Origin of Genetics: A Mendel Sourcebook,* San Francisco: Freeman, 1966, p. 110. In a recent article, Malcolm J. Kottler raised doubts about the plausibility of de Vries' claims regarding his independent discovery of the ratios: 'Hugo de Vries claimed that he had discovered Mendel's laws before he found Mendel's paper. De Vries' first ratios, published in 1897, for the second generation of hybrids (F_2) were $\frac{2}{3}:\frac{1}{3}$ and 80% : 20%. By 1900, both of these ratios had become 3 : 1. These changing ratios suggest that as late as 1897 de Vries had not discovered the laws, although he asserted, from 1900 on, that he had found the laws in 1896.' 'Hugo de Vries and the rediscovery of Mendel's laws', *Annals of Science,* vol. 36, 1979, p. 517.
6 Hugo de Vries, quoted in H. F. Roberts, *Plant Hybridization Before Mendel,* Princeton, N.J.: Princeton University Press, 1929, p. 328.
7 *Ibid.*
8 Bentley Glass, 'The long neglect of a scientific discovery: Mendel's laws of inheritance', in Johns Hopkins University History of Ideas Club, *Studies in Intellectual History,* Baltimore: Johns Hopkins University Press, 1953, p. 154.
9 *Ibid.*
10 It is ironic that Romanes borrowed Darwin's copy of Focke, and that neither apparently read the section on 'Leguminosae', for some of the pages in Darwin's copy had not been cut. Both Romanes and Darwin merely read the historical introduction. See Robert Olby, *The Origins of Mendelism,* London: Constance, 1966, p. 195.
11 Alexander Weinstein, 'How unknown was Mendel's paper?', *Journal of the History of Biology,* vol. 10, 1977, pp. 341–64, especially 341–2.

12 See Conway Zirkle, 'The role of Liberty Hyde Bailey and Hugo de Vries in the redis-covery of Mendelism', *Journal of the History of Biology,* vol. 1, 1968, pp. 205–18.

13 See A. E. Gaissinovitch, 'An early account of G. Mendel's work in Russia (I. F. Schmalhausen 1874)', in Milan Sosna (ed.), *G. Mendel Memorial Symposium 1865–1965,* Prague: Academia Publishing House of the Czechoslovak Academy of Science, 1966, pp. 39–40. Schmalhausen's report read, in part: 'His experiments and mathematical considerations in the second part of the work (*Befruchtungszellen der Hybriden*) lead him to conclusions which are basically similar to the theoretical considerations of Naudin', p. 40.

14 Weinstein, 'How unknown was Mendel's paper?' p. 343.

15 Glass, 'The long neglect of a scientific discovery', p. 149.

16 See Roberts, *Plant Hybridization Before Mendel,* p. 323.

17 L. C. Dunn, *A Short History of Genetics,* New York: McGraw Hill, 1965, p. 16.

18 See Roberts, *Plant Hybridization Before Mendel,* p. 337.

19 A. H. Sturtevant, *A History of Genetics,* New York: Harper and Row, 1965, p. 27.

20 See Roberts, *Plant Hybridization Before Mendel,* p. 339 (italics in original).

21 See Stern and Sherwood, *The Origin of Genetics,* p. 121 (italics in original).

22 See William B. Provine, *The Origins of Theoretical Population Genetics,* Chicago: University of Chicago Press, 1971, De Vries quoted on p. 68.

23 David L. Hull, Peter D. Tessner and Arthur M. Diamond, 'Planck's principle', *Science,* vol. 202, November 17, 1978, pp. 717–23, esp. pp. 720–1.

24 See Malcolm J. Kottler, 'Charles Darwin's biological species concept and theory of geographic speciation: the transmutation notebooks', *Annals of Science,* vol. 35, 1978, pp. 275–97. Also see Gerald L. Geison, 'Darwin and heredity: the evolution of his hypothesis of pangenesis', *Journal of the History of Medicine,* vol. 24, 1969, pp. 375–411.

25 See Fleeming Jenkin's discussion of swamping and the question of geological time in his review of *The Origin of Species* from the *North British Review* of June 1867, reproduced in David L. Hull, *Darwin and His Critics,* Cambridge, Mass.: Harvard University Press, 1973, pp. 302–44.

26 Provine, *The Origins of Theoretical Population Genetics,* pp. 9–10.

27 *Ibid.,* pp. 14ff.

28 William Bateson, *Materials for the Study of Variation,* New York: Macmillan, 1894. Also see Lindley Darden, 'William Bateson and the promise of Mendelism', *Journal of the History of Biology,* vol. 10, no. 1, 1977, pp. 87–106.

29 See Provine, *The Origins of Theoretical Population Genetics,* p. 56.

30 Gregor Mendel, 'Experiments on plant hybrids', translated by Evelyn Stern, in Stern and Sherwood, *The Origin of Genetics,* p. 35.

31 See Provine, *The Origins of Theoretical Population Genetics,* p. 28.

32 Pearson argued that organisms produce 'undifferentiated like organs' or 'homotypes' (e.g. blood cells, fish scales, body hair) which exhibit a degree of variability within the organism, but a degree smaller than that found for the race as a whole. Since the various ova and sperm cells are homotypic, they would unite to produce organisms the degree of variability between which would be no greater than the degree of variability in the homotypes of the parental organism. This model of heredity paid no special attention to the character of the germ cells and specifically the segregation which Mendel suggested occurred in them. See *ibid.,* pp. 58ff.

33 *Ibid.,* p. 61.

34 Weinstein, 'How unknown was Mendel's paper?', p. 360.

35 Eiseley, *Darwin's Century,* p. 206.

36 A. D. Darbishire, *Breeding and the Mendelian Controversy*, London: Cassell, 1911, p. 189. Cited by Robert Olby, 'Mendel no Mendelian?', *History of Science*, vol. 17, 1979, p. 53. For an even more outlandish account, cf. Wendt, *In Search of Adam*, p. 355. The correction of this misleading interpretation of the Mendel case was offered before Olby's work in a little known piece by Alan J. Bennett, 'Mendel's laws?', *School Science Review*, vol. 46, 1964, pp. 35–42. Bennett expressly anticipated the present author's hypothesis that Mendel was exploring hybridization as a key to the process of speciation.

37 Quoted in Vitezslav Orel, 'Response to Mendel's *Pisum* experiments in Brno since 1865', *Folia Mendeliana*, vol. 8, 1973, pp. 199–211, see p. 202.

38 *Ibid.*, p. 203.

39 *Ibid.*

40 *Ibid.*

41 *Ibid.*, p. 204.

42 *Ibid.*, p. 205.

43 See Ernst Mayr, 'The recent historiography of genetics', *Journal of the History of Biology*, vol. 6, 1973, pp. 125–54, see p. 140.

44 The Hawkweed plants which Mendel raised at the monastery in Brünn were apomictic plants: these plants appear to cross-fertilize while they in fact reproduce asexually – that is, without fertilization. Naegeli's focus on this plant to study heredity was unfortunate. Similarly the attention De Vries paid to the apparent new mutations in *Oenothera* was unwarranted; the peculiar behaviour of this species is produced by its balanced chromosome rings. (See *ibid.*, p. 137.)

45 Sir Ronald Fisher, 'Has Mendel's work been re-discovered?', pp. 139–72 in Stern and Sherwood, *The Origin of Genetics*, p. 171.

46 Gasking, 'Why was Mendel's work ignored?', p. 60.

47 *Ibid.*, p. 66.

48 Conway Zirkle, 'Gregor Mendel and his precursors', *Isis*, vol. 42, 1951, pp. 97–104, see p. 98.

49 *Ibid.*, p. 99.

50 Cf. *Ibid.*

51 *Ibid.*, p. 100.

52 *Ibid.*

53 Quoted in *ibid.*, p. 102.

54 Olby, *The Origins of Mendelism*, pp. 62–7.

55 Fisher, 'Has Mendel's work been re-discovered?', p. 164.

56 Mendel, 'Experiments in plant hybrids', p. 2.

57 See Orel, 'Response to Mendel's *Pisum* experiments', p. 201.

58 Mendel, 'Experiments in plant hybrids', p. 11.

59 *Ibid.*, p. 15.

60 *Ibid.*, pp. 41, 44, 47.

61 Quoted in Orel, 'Response to Mendel's *Pisum* experiments', p. 205.

62 Quoted in *ibid.*, p. 207.

63 See Roberts, *Plant Hybridization Before Mendel*, p. 168.

64 *Ibid.*, p. 180.

65 *Ibid.*, p. 81.

66 *Ibid.*, p. 82.

67 Quoted in *ibid.*, p. 155.

68 Quoted in *ibid.*, p. 96.

69 Mendel, 'Experiments in plant hybrids', p. 5.

70 *Ibid.*
71 *Ibid.*, p. 41 (italics in original).
72 *Ibid.*, p. 42 (italics in original).
73 *Ibid.*, pp. 45–6.
74 *Ibid.*, p. 47.
75 *Ibid.*, p. 22.
76 *Ibid.*, p. 37.
77 Quoted in Roberts, *Plant Hybridization Before Mendel*, p. 154.
78 *Ibid.*, p. 95.
79 Mendel, 'Experiments in plant hybrids', p. 37.
80 Zirkle, 'Mendel and his precursors', p. 101.
81 Contrast this view of Mendel with that suggested by Salvator Cannavo, *Nomic Inference*, The Hague: Martinus Nijhoff, 1974.
82 Cf. Hanna Pitkin, *Wittgenstein and Justice*, Los Angeles: University of California Press, 1972, pp. 70ff.
83 Walter B. Weimar, 'Science as rhetorical transaction: Toward a non-justificational conception of rhetoric', *Philosophy and Rhetoric*, vol. 10, no. 1, 1977, pp. 1–29, esp. pp. 1ff.
84 James D. Watson, *The Double Helix*, New York: Atheneum, 1968, p. 139.
85 Thomas S. Kuhn, 'Postscript 1969' to the second edition of *Structure of Scientific Revolutions*.
86 Lindley Darden, 'Reasoning in scientific change', p. 157.
87 Gasking, 'Why was Mendel's work ignored?', p. 69.
88 Conway Zirkle, 'Some oddities in the delayed discovery of Mendelism', *Journal of Heredity*, vol. 55, no. 2, 1964, p. 68.
89 Stern and Sherwood, *The Origin of Genetics*, pp. x–xi.
90 Arthur Koestler, *The Case of the Midwife Toad*, London: Hutchinson, 1971.
91 Joseph Weiner, *The Piltdown Forgery*, London: Oxford University Press, 1955; Ronald Millar, *The Piltdown Men*, Letchworth: Victor Gollancz, 1972.

7 *Perspective, reflexivity and the apparent objectivity of discovery*

1 Samuel Eliot Morison, *The European Discovery of America*, New York: Oxford University Press, 1971, p. 7.
2 Thomas S. Kuhn, 'Historical structures of scientific discoveries', *Science*, vol. 136, 1962, p. 762.
3 The doctrine of the spheroid shape of the earth had been established by both classical and biblical sources. Aristotle's observation of the circular shape of the earth cast on the moon during an eclipse had established this. The Pythagoreans had postulated that the shape of the earth was a sphere because of the perfection of the shape. Medieval scholastics had found corroboration of these views in the Scriptures. Isaiah xl. 22: 'It is he that sitteth on the circle of the earth.' Columbus based his earliest projections on the apocryphal book of *Esdras*.
4 Hans Selye, *From Dream to Discovery*, New York: McGraw Hill, 1964, p. 88.
5 Melvin Pollner, 'Mundane reasoning', *Philosophy of the Social Sciences*, vol. 4, 1974, pp. 35–54.
6 Cf. Aaron Cicourel, 'The acquisition of social structure', pp. 136–68 in *Understanding Everyday Life*, edited by Jack Douglas, Chicago: Aldine, 1970, pp. 147–8.
7 This is discussed by Melvin Pollner in reference to labelling accounts of deviance. A

sociological account of deviance cannot be satisfied with the simple labelling account, for this does not show why deviance is so compelling as a moral fact for the member of society. Melvin Pollner, 'Sociological and common-sense models of the labelling process', pp. 27–41 in *Ethnomethodology*, edited by Roy Turner, Harmondsworth: Penguin, 1974.

8 Pollner, 'Mundane reasoning', n. 3.

9 Harold Garfinkel, 'Studies of the routine grounds of everyday activities', in *Studies in Ethnomethodology*, Englewood Cliffs: Prentice-Hall, 1967, pp. 35–75.

10 Pollner, 'Mundane reasoning', p. 36.

11 Morison, *The European Discovery of America*, p. 3.

12 Having arrived at Guanahani, Columbus sailed to the other islands following the directions given to him by the natives from island to island. The point here is that the further explorations were not guided by induction, serendipity, or insight, but by the use of the common stock of knowledge of the native peoples. This was also what guided Darwin in his 'discovery' of the variations among the Galapagos Islands. The differences between the tortoises was pointed out to him by local sailors.

13 The distance was arrived at by Columbus using an erroneous set of calculations. Basing his estimation on the prophet *Esdras*, he calculated: '(i) the earth is round, (ii) the distance by land between the edge of the west (Spain) and the edge of the east ("India" i.e. Asia) is very long; (iii) the distance by sea between Spain and "India" is very short; (iv) the length of the degree is 56⅔ miles. These miles were not Arabic (1,975.5 meters) which would have made the figure remarkably accurate, but Italian (1,477.5 meters) which made his equator about one quarter too small. Columbus calculated that the land distance between Spain and "India" was 282 degrees; he was therefore left with only 78 degrees for the sea distance, which he further reduced by his method of reckoning the degree. The outcome of all these errors was that "India" would be about 3,900 miles from the Canaries; i.e. more or less where America happens to be.' Salvador de Madariaga, 'Christopher Columbus', pp. 937–42 in *Encyclopaedia Britannica*, vol. 4, fifteenth edition, 1974, p. 938.

14 It was characteristic of those institutional responses that the geographic discoveries had dramatic political and economic overtones. New-found land became the crown's realty and its indigenous peoples became the crown's subjects. Also, the natural wealth could be sold by the crown via licences or charters to joint stock ventures, as was the case in New France and Hudson Bay. However, all such geographic discoveries had a scientific interest in cartography. This superseded the other claims, for only if the land was assumed to be undiscovered could it be claimed. The political infusion of other kinds of discovery and invention has not been uncommon. This is discussed by Bernard Barber in *Science and the Social Order*, New York: 1952, p. 261.

15 A thematically identical letter was written by Darwin to his wife in 1842; cf. Charles Darwin, *Autobiography and Selected Letters*, edited by Francis Darwin (1892), New York: Dover 1958, pp. 180–2.

16 John Boyd Thacker, *Christopher Columbus*, vol. 2, New York: AMS Press, (1902) 1967, pp. 72–3.

17 For a compelling discussion of the reflexive 'exhibition' of research findings as playbacks of research scenes, see Lawrence Wieder, 'Telling the code', in Turner, *Ethnomethodology*, pp. 170–2.

18 However, *that* we exhibit the same opinion is a folk act, and *not* a sociological description. Yet even our sociological study presupposes the folkways and relies on them for its accomplishments. In other words, our practices for uncovering the so-

cial basis for discovery are already apprised of these folkways as resources. Our usage is unavoidably part of the events we are trying to describe.

19 Edmundo O'Gorman, *The Invention of America,* Bloomington: Indiana University Press, 1961, pp. 127–30.

20 *Ibid.,* 'The process of the invention of America', pp. 73–124, esp. pp. 115ff.

21 Thomas S. Kuhn, *The Structure of Scientific Revolutions,* Chicago: University of Chicago Press, 1970, p. 53.

22 *Ibid.,* p. 53.

23 *Ibid.,* p. 54.

24 Dephlogisticated air was air bereft of phlogiston, an imagined element which was believed to transfer out of combustible materials during burning.

25 Kuhn, *Structure of Scientific Revolutions,* p. 55.

26 Denis I. Duveen, 'Lavoisier', in *Encyclopaedia Britannica,* vol. 10, fifteenth edition, 1974, pp. 713–14.

27 James D. Watson, *The Double Helix,* New York: Atheneum, 1968.

28 Cf. Uno Bocklund, 'A lost letter from Scheele to Lavoisier', *Lychnos,* vol. 39, 1957–8, pp. 39–62.

29 Kuhn, *Structure of Scientific Revolutions,* p. 55.

30 *Ibid.*

31 *Ibid.* (italics added).

32 Joseph Weiner, *The Piltdown Forgery,* London: Oxford University Press, 1955, p. 1.

33 Ronald Millar, *The Piltdown Men,* Letchworth: Victor Gollancz, 1972, p. 130.

34 Weiner, *The Piltdown Forgery,* p. 44.

35 Compare with Garfinkel's conclusion to a study of Los Angeles' Suicide Prevention Center: '*Whatsoever* SPC members are faced with must serve as the precedent with which to read the remains so as to see how the society could have operated to have produced what it is that the inquiry has "in the end", "in the final analysis" and "in *any* case" '; in Turner, *Ethnomethodology,* p. 101. The same general point is indicated by the discussion of local settings as an *occasioned corpus* in D. Zimmerman and M. Pollner, 'The everyday world as a phenomenon' pp. 80–103 in *Understanding Everyday Life,* edited by Jack Douglas, Chicago: Aldine, 1970.

36 Woodward, quoted in Weiner, *The Piltdown Forgery,* p. 9.

37 Indeed recent tests have put the date at no more than 500 years before the present.

38 Tests measured the concentration of fluorine, nitrogen, organic carbon, organic water, etc.; each element is present in different proportions in recent as opposed to ancient remains. Hence, fossils have accumulated more fluorine and have lost more nitrogen than modern deposits.

39 After he died, it was brought to light that Dawson had a nearly complete skeleton of recent origin taken from the Barcombe Mills area. It was dyed and preserved identically to the Piltdown remains. Weiner, *The Piltdown Forgery,* pp. 151–2.

40 As sociologists, we must resist the temptation to jump to conclusions about 'what really happened', since whatever occurred is a function of the methods used to gloss or interrogate a scene; *the use* of such methods is the phenomenon of paramount interest in this study. However, the fact that there do exist different accounts or formulations of the events is nonetheless instructive, since it undermines our attachment to any particular account as a definitive one which could be used authoritatively to 'mop up' the details. Such a move would adumbrate the process by which the 'facts' are constituted by such methods.

41 Millar, *The Piltdown Men,* p. 237.

42 Quoted in *ibid.,* p. 207.

43 Cf. Nicholas Wade, 'Voice from the dead names new suspect for Piltdown hoax',

Science, vol. 202, December 8, 1978, p. 1062. Also see S. L. Washburn, 'The Pilt-down hoax: Piltdown 2', letter to the editor, *Science,* vol. 203, March 9, 1979, pp. 955–8.

8 Folk reasoning in theories about scientific discovery

1 Paul Forman, 'The discovery of the diffraction of X-rays by crystals: a critique of the myths', *Archive for History of Exact Sciences,* vol. 6, 1969, pp. 38–71. The suspicion of myth-making surrounding scientific personalities is also given a lively discussion in S. C. Gilfillan, 'Who invented it?', *Scientific Monthly,* vol. 25, 1927, pp. 529–34. Gilfillan points out that for many prominent discoveries every country has its own roster of scientific heroes whose accomplishments are exalted as part of the national heritage. As Gilfillan implies, this is an important consideration in evaluating the reliability of lists of multiple discoveries.

2 Robert Olby, 'Mendel no Mendelian?', *History of Science,* vol. 17, 1979, pp. 53–72. In terms of dates, Olby's work anticipated my own ideas by several months. In addition, his breadth of understanding of the historical context and his technical proficiency with Mendel's analysis dwarf my contributions. These are some of Olby's more perspicuous conclusions: 'How far did [Mendel] go in his conception of a particulate theory of heredity? First it is evident that he did not conceive of pairs of elements in the cell representing and determining the pairs of contrasted characters. If he had this conception he would have allowed a separation *between like members of such pairs as well as between unlike members.* His statement that "only the differing elements are mutually exclusive" is in conflict with classical Mendelian genetics' (p. 66); 'Mendel's overriding concern was with the role of hybrids in the genesis of new species' (p. 67); 'These results lead one to conclude that the problem of the long neglect of the *Versuche* is to a large extent a pseudo-problem' (p. 56); 'if we arbitrarily define a Mendelian as one who subscribes explicitly to the existence of a finite number of hereditary elements which in the simplest case is two per hereditary trait, only one of which may enter one germ cell, then Mendel was clearly no Mendelian' (p. 68).

I was led to some of the same conclusions by reading *between* the lines of Olby's earlier analysis in *The Origins of Mendelism,* London: Constance, 1966. Another major impetus to my thought was the overbearing implausibility of the long neglect explanations.

3 Cf. chapter 7.

4 Nicholas Wade, 'I.Q. and heredity: suspicion of fraud beclouds classic experiment', *Science,* vol. 194, November 20, 1976, pp. 916–19.

5 Bernard Norton, 'A "fashionable fallacy" defended', *New Scientist,* vol. 78, April 27, 1978, pp. 223–5.

6 Stephen J. Gould, 'Morton's ranking of races by cranial capacity', *Science,* vol. 200, May 5, 1978, pp. 503–9.

7 A. L. Kroeber, 'The superorganic', *American Anthropologist,* vol. 19, no. 2, 1917, p. 200.

8 *Ibid.,* pp. 198–9.

9 Robert K. Merton, *The Sociology of Science, Theoretical and Empirical Investigations,* Chicago: University of Chicago Press, 1973, p. 356.

10 Leslie White, *The Science of Culture,* second edition, New York: Farrar, Straus and Giroux, 1969.

11 *Ibid.,* p. 214.

12 Jacques Hadamard, *The Psychology of Invention in the Mathematical Field*, Princeton: Princeton University Press, 1949, p. 132.

13 Merton, *The Sociology of Science*, p. 357.

14 William F. Ogburn, *Social Change*, New York: Heubsch, 1922, p. 86.

15 Merton, *The Sociology of Science*, p. 380.

16 Cf. *ibid.*, pp. 337, 365.

17 Edward Sapir, 'Do we need a superorganic?', *American Anthropologist*, vol. 19, 1917, pp. 442–3.

18 Kroeber, 'The superorganic', p. 197.

19 The list of multiples was published in two places: as an appendix to an article authored by William F. Ogburn and Dorothy Thomas, 'Are inventions inevitable? A note on social evolution', *Political Science Quarterly*, vol. 37, no. 1, 1922, pp. 83–98; and as an appendix to a chapter in Ogburn's 1922 book on *Social Change*. In both places, the footnotes added to the appendix explaining problems in the list appear to have been composed by Dorothy Thomas. The list is too long to include here, and readers are encouraged to consult the original sources. It is argued in the text that over one-half of the 1922 list of multiple discoveries could be discarded on the assumption that duplications resulted from a failure of communication among the concerned parties. This figure was determined in the following way: all the entries were examined to identify those discoveries which were recorded in *separate years*. There were eighty-five items. The numbers here are those given to the entries by Dorothy Thomas in the original list. Items appearing in different years ($n = 85$):

    2, 3, 4, 8, 9, 15, 17, 18, 19, 20, 22, 23, 24, 25, 28, 29, 30, 33, 34, 35, 36, 37, 39, 40, 45, 48, 49, 51, 52, 54, 56, 58, 59, 62, 63, 65, 66, 67, 69, 70, 71, 74, 75, 76, 79, 80, 82, 83, 84, 85, 86, 89, 91, 92, 95, 96, 97, 98, 99, 100, 105, 108, 109, 111, 115, 116, 117, 118, 125, 127, 128, 129, 130, 132, 133, 134, 136, 138, 139, 142, 143, 144, 145, 146, 148.

If we were to identify only those items which were a year apart (for example, 1774 and 1775) we could *remove* just ten items from this list (51, 52, 62, 85, 86, 96, 117, 133, 139, 146); but the majority of the items are separated by more than a year. Removing these we see that about one half the list can be challenged on the basis of the assumption in the text. Furthermore, the remainder of the list which is unchallenged has other problems. Ten items are undated or record dates for only one of the entries. Several are marked by 'claimed by' (67, 108 and 109), indicating Thomas' suspicions. And lastly a number of items are dated ambiguously (21, 41, 53, 68, 87, 101, 103, 107, 114, 119, 126, 140).

20 Ogburn and Thomas, 'Are inventions inevitable?', p. 93.

21 Lindley Darden, 'Williams Bateson and the promise of Mendelism', *Journal of the History of Biology*, vol. 10, no. 1, 1977, p. 92.

22 H. F. Roberts, *Plant Hybridization Before Mendel*, Princeton, N.J.: Princeton University Press, 1929, pp. 343ff.

23 Thomas S. Kuhn, *The Structure of Scientific Revolutions*, second edition, Chicago: University of Chicago Press, 1970, p. 53.

24 Francis Galton, *Hereditary Genius* (1869), Gloucester, Mass.: Meridian Books, 1962.

25 Thomas S. Kuhn, 'Historical structures of scientific discoveries', *Science*, vol. 136, 1962, p. 763.

26 Norwood Russell Hanson, 'Notes toward a logic of discovery', pp. 42–65 in *Perspectives on Peirce*, edited by R. J. Bernstein, New Haven: Yale University Press, 1965.

27 Norwood Russell Hanson, 'The logic of discovery', *Journal of Philosophy*, vol. 55, no. 25, 1958, p. 1083.

28 R. J. Blackwell, *Discovery in the Physical Sciences*, Notre Dame: Notre Dame University Press, 1969, p. 19.
29 A. Koestler, *The Act of Creation*, London: Hutchinson, 1964.
30 Henri Poincaré, *Science and Method*, New York: Scribner, 1914.
31 Melvin Pollner, 'Mundane reasoning', *Philosophy of the Social Sciences*, vol. 4, 1974, pp. 35–54; Pollner, 'The very coinage of your brain', *Philosophy of the Social Sciences*, vol. 5, 1975, pp. 411–30.
32 Alfred Schutz, *The Collected Papers*, vol. 1: *The Problem of Social Reality*, The Hague: Martinus Nijhoff, 1962.
33 Pollner, 'Mundane reasoning', pp. 49ff.
34 Oscar Handlin, 'Ambivalence in the popular response to science', pp. 253–68 in *The Sociology of Science*, edited by Barry Barnes, Harmondsworth: Penguin, 1972.
35 Merton, *The Sociology of Science*, pp. 352–6.
36 H. Garfinkel and H. Sacks, 'On formal structures of practical activities', pp. 338–66 in *Theoretical Sociology*, edited by John C. McKinney and Edward Tiryakian, New York: Appleton, Century, Crofts, 1970.
37 Bennetta Jules-Rosette, 'The veil of objectivity', *American Anthropologist*, vol. 80, 1978, pp. 549–70.
38 James L. Heap, 'What are sense-making practices?', *Sociological Inquiry*, vol. 46, no. 2, 1976, pp. 107–15.
39 Felix Kaufman, *The Methodology of the Social Sciences*, New York: Oxford University Press, 1944, pp. 33ff.
40 Alfred Schutz, *Phenomenology of the Social World* (1932), translated by George Walsh and Frederick Lehnert, Evanston: Northwestern University Press, 1967.
41 Koestler, *The Act of Creation*, pp. 87ff.
42 Poincaré, *Science and Method*, pp. 55ff.

9 *Some sociological features of discovery*

1 For the purposes of sociological analysis, it does not matter in the end what the 'real facts' are regarding novelty, etc. These matters are only of note as considerations of members of society. H. Garfinkel and H. Sacks, ('On formal structures of practical activities', pp. 338–66 in *Theoretical Sociology*, edited by John C. McKinney and Edward A. Tiryakian, New York: Appleton, Century, Crofts, 1970, p. 343) also describe a position of *indifference* to the ultimate significance of members of society's phenomena; cf. n. 2. This position is essential methodologically if the observer is to have access to folk phenomena *as* folk phenomena.
2 These are examined closely in James L. Heap, 'What are sense-making practices?', *Sociological Inquiry*, vol. 46, no. 2, 1976, pp. 107–15; Heap's analysis concentrates for the most part on Cicourel's position in 'The acquisition of social structure', pp. 136–68 in *Understanding Everyday Life*, edited by Jack D. Douglas, Chicago: Aldine, 1970.
3 Aaron Cicourel, 'Basic and normative rules', pp. 4–45 in *Recent Sociology, Number Two*, edited by H. P. Dreitzel, New York: Macmillan, 1970.
4 Cf. Harold Garfinkel, *Studies in Ethnomethodology*, Englewood Cliffs: Prentice-Hall, 1967, pp. 76–103.
5 'Normal form' is discussed by Cicourel, 'The acquisition of social structure'; a normal form is a typified scheme of interpretation used by members of society to organize the particulars of a scene, especially when the definition of the situation or the normative order is ambiguous, or is undergoing a change. Notes Cicourel (p. 148): 'On

occasions when the reciprocity of perspectives is in doubt, efforts will be made by both speaker and hearer to normalize the presumed discrepancies.' This attempt to normalize the reciprocity of perspectives involves the offering of formulations about what the meaning or significance of certain scenes could be.

6 Some rather good cognitive analyses of scientific settings are found in the following: Bruno Latour and Steve Woolgar, *Laboratory Life: The Social Construction of Scientific Facts,* Beverly Hills and London: Sage, 1979; Karen D. Knorr, 'Producing and reproducing knowledge: descriptive or reconstructive', *Social Science Information,* vol. 16, no. 6, 1977, pp. 669–96. Also see David O. Edge and Michael J. Mulkay, *Astronomy Transformed, The Emergence of Radio Astronomy in Britain,* New York: Wiley Interscience, 1976; H. M. Collins, 'The seven sexes: a study in the sociology of a phenomenon, or the replication of experiments in physics', *Sociology,* vol. 9, 1975, pp. 205–24.

7 As seen in this instance, the normal form of an event is not an inherent order or organization, but is an imputed or assumed order.

8 Joseph Weiner, *The Piltdown Forgery,* London: Oxford University Press, pp. 176ff.

9 Ronald Millar, *The Piltdown Men,* Letchworth: Victor Gollancz, 1972, pp. 223–4.

10 The notion of performatives was first identified by J. L. Austin, 'Performative utterances', in *Philsophical Papers,* edited by J. O. Urmson and G. L. Warnock, Oxford: Oxford University Press, 1961. Further investigations of performatives have appeared in the writings of Roy Turner, 'Words, utterances and activities', pp. 169–87 in *Understanding Everyday Life,* edited by Jack Douglas, Chicago: Aldine, 1970; and Hanna Pitkin, *Wittgenstein and Justice,* Los Angeles: University of California Press, 1972.

11 Pitkin, *Wittgenstein and Justice,* pp. 280ff.

12 Edward W. Constant II, 'On the diversity and co-evolution of technological multiples', *Social Studies of Science,* vol. 8, 1978, pp. 183–210. See p. 183.

13 *Ibid.*

14 *Ibid.,* p. 197.

15 Hannah Gay, 'The asymmetric carbon atom', *Studies in the History and Philosophy of Science,* vol. 9, 1978, pp. 207–38.

16 *Ibid.,* p. 222.

17 *Ibid.,* p. 237. There is of course something of a paradox in this position as has been indicated by Ron Westrum: 'Merton argues that there is a common intellectual state of society which leads to independent simultaneous inventions. Yet of what does this independence consist? It consists of a lack of common intellectual elements – that is to say, lack of communication between the inventors. But, on the other hand, it is the fact of having a common intellectual heritage, Merton argues, which permits simultaneous invention in the first place. To the extent that they share a common intellectual culture, then, they are less independent. One cannot both increase their common heritage and their independence at the same time!' See the discussion, 'The notion of independent simultaneous invention or discovery', *Social Studies of Science,* vol. 9, 1979, pp. 508ff.

18 Frank Brescia and Pietro Mangiaracina, 'Conformational analysis, 1869', *Journal of Chemical Education,* vol. 53, no. 1, 1976, pp. 32–3.

19 Yehuda Elkana, *The Discovery of the Conservation of Energy,* Cambridge, Mass.: Harvard University Press, 1974.

20 David A. Hounshell, 'Elisha Gray and the telephone: on the disadvantages of being an expert', *Technology and Culture,* vol. 16, 1975, pp. 133–61.

21 Elizabeth N. Shor, *The Fossil Feud Between E. D. Cope and O. C. Marsh,* New York: Exposition Press, 1974.

22 Much of Ignaz Semmelweis' book relating his discovery of the source of child-bed fever and his development of a method to prevent it is contained in W. J. Sinclair, *Semmelweis: His Life and His Doctrine*, Manchester: Manchester University Press, 1909.

23 Carl G. Hempel, *Philosophy of Natural Science*, Englewood Cliffs: Prentice-Hall, 1966, pp. 3ff.

24 G. N. Gilbert, 'The transformation of research findings into scientific knowledge', *Social Studies of Science*, vol. 6, 1976, pp. 281–306.

25 Of related interest here is a study by James Stephens, *Francis Bacon and the Style of Science*, Chicago: University of Chicago Press, 1975. According to Stephens, Bacon argued that scientists must utilize their style to control the imagination of their readers so as to gain 'quiet entry' into the readers' minds. 'The duty and office of Rhetoric, if it be looked into', Bacon says, 'is no other than to apply and recommend the dictates of reason to imagination, in order to excite the appetite and will' (p. 57).

A more prosaic discussion of these issues is found in John Ziman, *Reliable Knowledge*, Cambridge: Cambridge University Press, 1978; see 'Unambiguous communication', pp. 11–41, and esp. p. 12 for a discussion of Ryle's 'didactic discourse'.

Drawing on the formal machinery of conversational analysis, certain recent authors have begun to devise methods to conduct very fine grained analyses of what Bacon saw as the rhetorical structure of scientific texts. Cf. Digby Anderson, 'Some organizational features in the local production of a plausible text', *Philosophy of Social Sciences*, vol. 8, 1978, pp. 113–35; Steve Woolgar, 'Discovery: logic and sequence in a scientific text', in *Sociology of the Sciences Yearbook*, vol. 5: *The Social Process of Scientific Investigation*, Dordrecht: Reidel, 1981, forthcoming; G. N. Gilbert and Michael J. Mulkay, 'Contexts of scientific discourse: social accounting in experimental papers', *The Social Process of Scientific Investigation*, edited by K. D. Knorr *et al.*, Dordrecht: Reidel, 1980, pp. 269–94.

In a similar vein some recent work by Karen D. Knorr deals with many of the same issues. See 'From scenes to scripts: on the relationship between laboratory research and published paper in science', Research Memorandum no. 132, August, 1978, Institute for Advanced Studies, Vienna; 'Contextuality and indexicality of organizational action: toward a transorganizational theory of organizations', *Social Science Information*, vol. 18, no. 1, 1979, pp. 79–101; 'Producing and reproducing knowledge'; 'Tinkering toward success: prelude to a theory of scientific practice', *Theory and Society*, vol. 8, 1979, pp. 347–76.

26 A recent study by Wanner, Lewis and Gregorio compares the research productivity (i.e. books and articles) of academics in the sciences, the social sciences and the humanities. The authors indicate that when we control for the individual characteristics of the scientists and their backgrounds, one of the important predictors of innovative behaviour is the nature of the disciplines themselves. 'Our evidence suggests that the decisive edge that physical and biological scientists enjoy over social scientists and humanists in article productivity is largely the result of a favorable disciplinary milieu while the lower rate of productivity among humanists is decisively determined by their attributes.' In a large sample of U.S. academics, the average number of articles published was 13.25, 9.43 and 5.64 for the sciences, social sciences and humanities respectively. The average number of books published for the same three groups was 0.89, 1.8 and 1.44 respectively. Of course, one of the assumptions of this work is that every article and book is indicative of innovative behaviour. Few would take this for granted, including the authors. Nonetheless, this is one of the only studies which compares the output of academics across disciplines and which examines their productivity controlling for both academic and

background characteristics. See Richard A. Wanner, Lionel S. Lewis and David I. Gregorio, 'Research productivity in academia: a comparative study of the sciences, social sciences and humanities', Sociology of Science Session, Annual Meetings of the American Sociological Association, New York City, August, 1980, forthcoming in *Sociology of Education,* vol. 54, 1981.

Bibliography

Anscombe, G. E. M. 'Causality and determination', pp. 63–81 in *Causation and Conditionals,* edited by E. Sosa. Oxford: Oxford University Press, 1975

Aristotle. *The Works of Aristotle,* translated under the editorship of W. D. Ross. Oxford: Clarendon Press, 1928

Austin, J. L. *Philosophical Papers,* edited by J. O. Urmson and G. L. Warnock. Oxford: Oxford University Press, 1961
  How To Do Things With Words, edited by J. O. Urmson. New York: Oxford University Press, 1965

Barber, Bernard. *Science and the Social Order.* New York: Macmillan, 1952
  'Resistance by scientists to scientific discovery', *Science,* vol. 134, Sept. 1, 1961, pp. 596–602
  and Walter Hirsch (eds.). *The Sociology of Science.* New York: Free Press, 1962

Barnes, Barry (ed.). *The Sociology of Science.* Harmondsworth: Penguin, 1972
  Scientific Knowledge and Sociological Theory, London: Routledge and Kegan Paul, 1974
  and S. Shapin (eds.). *Natural Order: Historical Studies of Scientific Culture.* Beverly Hills and London: Sage, 1979

Baumrim, Bernard (ed.). *Philosophy of Science: The Delaware Seminar* (1961–2), vol. 1. New York: Wiley Interscience, 1963

Bell, E. T. *Men of Mathematics.* New York: Simon and Schuster, 1937

Blackwell, Richard J. *Discovery in the Physical Sciences.* Notre Dame: Notre Dame University Press, 1969

Blalock, Jr, H. M. *Theory Construction.* Englewood Cliffs: Prentice-Hall, 1969

Bloor, David. 'Wittgenstein and Mannheim on the sociology of mathematics', *Studies in the History and Philosophy of Science,* vol. 4, no. 2, 1973, pp. 173–91
  Knowledge and Social Imagery. London: Routledge and Kegan Paul, 1976
  'Polyhedra and the abominations of Leviticus', *British Journal for the History of Science,* vol. 11, no. 39, 1978, pp. 245–72

Bocklund, Uno. 'A lost letter from Scheele to Lavoisier', *Lychnos,* vol. 39, 1957–8, pp. 39–62

Boring, E. G. *History, Psychology and Science, Selected Papers,* edited by R. J. Watson and D. T. Campbell, New York: Wiley, 1963

Braithwaite, Richard. *Scientific Explanation.* Cambridge: Cambridge University Press, 1953

Butler, Samuel. *Erewhon; or, Over the Range,* revised edition. London: A. C. Filfield, 1908

Butterfield, Herbert. *The Origins of Modern Science.* London: G. Bell, 1949

Cannavo, Salvator. *Nomic Inference.* The Hague: Martinus Nijhoff, 1974

Cicourel, Aaron V. 'Basic and normative rules', pp. 4–45 in *Recent Sociology, Number Two,* edited by H. P. Dreitzel. New York: Macmillan, 1970

'The acquisition of social structure', pp. 136–68 in *Understanding Everyday Life,* edited by Jack D. Douglas. Chicago: Aldine, 1970

et al. Language Use and School Performance. New York: Academic Press, 1974

Cole, Jonathan and Stephen Cole. *Social Stratification in Science.* Chicago: University of Chicago Press, 1973

Collins, H. M. 'The seven sexes: a study in the sociology of a phenomenon, or the replication of experiment in physics', *Sociology,* vol. 9, 1975, pp. 205–24

Constant II, Edward W. 'On the diversity and co-evolution of technological multiples', *Social Studies of Science,* vol. 8, 1978, pp. 183–210

Cooley, Charles Horton. 'Genius, fame and the comparison of races', *Annals of the American Academy of Political and Social Science,* vol. 9, 1897, pp. 1–42

Crane, Diana. *Invisible Colleges.* Chicago: University of Chicago Press, 1972

Darden, Lindley. 'Reasoning in scientific change: Charles Darwin, Hugo de Vries, and the discovery of segregation', *Studies in the History and Philosophy of Science,* vol. 7, no. 2, 1976, pp. 126–69

'William Bateson and the promise of Mendelism', *Journal of the History of Biology,* vol. 10, no. 1, 1977, pp. 87–106

Darwin, Charles. *The Origin of Species.* London: John Murray, 1859

The Variation of Plants and Animals Under Domestication. London: John Murray, 1868

Autobiography and Selected Letters, edited by Francis Darwin (1892). New York: Dover, 1958

Descartes, René. *Discourse on Method* (1637). Translated by L. J. Lafleur. Indianapolis: Bobbs-Merrill, 1950

Desmond, Adrian. *The Hot-Blooded Dinosaurs.* London: Blond and Briggs, 1975

Dolby, R. G. A. 'The sociology of knowledge in natural science', pp. 309–320 in *The Sociology of Science,* edited by Barry Barnes. Harmondsworth: Penguin, 1972

Douglas, Jack D. (ed.). *Understanding Everyday Life.* Chicago: Aldine, 1970

Drake, Stillman. *The Discoveries and Opinions of Galileo.* New York: Anchor Books, 1957

Duveen, Denis I. 'Lavoisier', pp. 713–14 in *Encyclopaedia Britannica,* vol. 10, fifteenth edition. Chicago: H. H. Benton, 1974

Edge, David O. 'Quantitative measures of communication in science: a critical review', *History of Science,* vol. 17, 1979, pp. 102–34

and Michael J. Mulkay *Astronomy Transformed. The Emergence of Radio Astronomy in Britain.* New York: Wiley Interscience, 1976

Eiseley, Loren. *Darwin's Century*. New York: Anchor Books, 1961

Elkana, Yehuda. *The Discovery of the Conservation of Energy*. Cambridge, Mass.: Harvard University Press, 1974

Ellegard, Alvar. *Darwin and the General Reader*. Göthenburg: Goteborgs Universitets Arsskrift, 1958

Evans-Pritchard, E. E. *Witchcraft, Oracles and Magic Among the Azande*, abridged edition. Oxford: Clarendon Press, 1976

Feigl, H. and G. Maxwell (eds.). *Current Issues in the Philosophy of Science*. New York: Holt, Rinehart and Winston, 1961

Fell, Barry. *America B.C.* New York: Quadrangle Books, 1976

Feuer, Lewis S. *Einstein and the Generations of Science*. New York: Basic Books, 1974

'Historical method in the sociology of science: the pitfalls of a polemicist', *Philosophy of the Social Sciences*, vol. 7, 1977, pp. 255–61

Feyerabend, Paul K. *Against Method*. London: New Left Books, 1975

Fisher, Sir Ronald A. 'Has Mendel's work been re-discovered?' (1936), pp. 139–72 in *The Origin of Genetics: A Mendel Sourcebook*, edited by Curt Stern and Eva Sherwood. San Francisco: Freeman, 1966

Forman, Paul. 'The discovery of the diffraction of X-rays by crystals: a critique of the myths', *Archive for History of Exact Sciences*, vol. 6, 1969, pp. 38–71

Freud, Sigmund. *Leonardo da Vinci*. Translated by Alan Tyson. New York: Norton, 1964

Galton, Francis. *Hereditary Genius* (1869). Gloucester, Mass.: Meridian Books, 1962

Natural Inheritance. London: Macmillan, 1889

Finger Prints. London: Macmillan, 1892

Garfinkel, Harold. *Studies in Ethnomethodology*. Englewood Cliffs: Prentice-Hall, 1967

Gasking, Elizabeth. 'Why was Mendel's work ignored?', *Journal for the History of Ideas*, vol. 20, 1959, pp. 60–84

Gaston, Jerry. *Originality and Competition in Science*. Chicago: University of Chicago Press, 1973

Gay, Hannah. 'The asymmetric carbon atom', *Studies in the History and Philosophy of Science*, vol. 9, 1978, pp. 207–38

Ghiselin, Brewster (ed.). *The Creative Process*. Los Angeles: University of California Press, 1952

Gilbert, G. N. 'The transformation of research findings into scientific knowledge', *Social Studies of Science*, vol. 6, 1976, pp. 281–306

and Michael J. Mulkay. 'Accounting for error: how scientists construct their social world when they account for correct and incorrect belief', *Sociology*, forthcoming

'Contexts of scientific discourse: social accounting in experimental papers', in *The Social Process of Scientific Investigation*, edited by K. D. Knorr *et al*. Dordrecht: Reidel, 1980, pp. 269–94.

Gilfillan, S. C. 'Who invented it?', *Scientific Monthly*, vol. 25, 1927, pp. 529–34

Glass, Bentley. 'The long neglect of a scientific discovery: Mendel's laws of inheritance', pp. 148–60 in Johns Hopkins University History of Ideas Club, *Studies in Intellectual History*. Baltimore: Johns Hopkins University Press, 1953

Goffman, Erving. *Frame Analysis*. New York: Colophon Books, 1974

Gould, Stephen J. 'Morton's ranking of races by cranial capacity', *Science,* vol. 200, May 5, 1978, pp. 503–9

Hadamard, Jacques. *The Psychology of Invention in the Mathematical Field*. Princeton: Princeton University Press, 1949

Hagstrom, Warren O. *The Scientific Community*. New York: Basic Books, 1965

Handlin, Oscar. 'Ambivalence in the popular response to Science', pp. 253–68 in *The Sociology of Science,* edited by Barry Barnes. Harmondsworth: Penguin, 1972

Hanson, Norwood Russell. 'The logic of discovery', *Journal of Philosophy,* vol. 55, no. 25, 1958, pp. 1073–89

  Patterns of Discovery. Cambridge: Cambridge University Press, 1958

  'Is there a logic of discovery?', pp. 20–34 in *Current Issues in the Philosophy of Science,* edited by H. Feigl and G. Maxwell. New York: Holt, Rinehart and Winston, 1961

  'Retroductive inference', pp. 21–37 in *Philosophy of Science: The Delaware Seminar,* vol. 1, edited by Bernard Baumrim. New York: Wiley Interscience, 1963

  'Notes toward a logic of discovery', pp. 42–65 in *Perspectives on Peirce,* edited by R. J. Bernstein. New Haven: Yale University Press, 1965

  'An anatomy of discovery', *Journal of Philosophy,* vol. 64, no. 11, 1967, pp. 321–52

Hardin, Garrett. *Nature and Man's Fate*. New York: Holt, Rinehart and Winston, 1959

Heap, James L. 'What are sense-making practices?', *Sociological Inquiry,* vol. 46, no. 2, 1976, pp. 107–15

Heidegger, Martin. *What Is a Thing?* Chicago: Henry Regnery, 1970

Hempel, Carl G. *Philosophy of Natural Science*. Englewood Cliffs: Prentice-Hall, 1966

Holton, Gerald. *Thematic Origins of Scientific Thought*. Cambridge, Mass.: Harvard University Press, 1973

  The Scientific Imagination. New York: Cambridge University Press, 1978

Hounshell, David A. 'Elisha Gray and the telephone: on the disadvantages of being an expert', *Technology and Culture,* vol. 16, 1975, pp. 133–61

Hull, David L. *Darwin and His Critics*. Cambridge, Mass.: Harvard University Press, 1973

  'Altruism in science', *Animal Behavior,* vol. 26, 1978, pp. 685–97

  Peter D. Tessner and Arthur M. Diamond. 'Planck's principle', *Science,* vol. 202, November 17, 1978, pp. 717–23

Huxley, Aldous. *Brave New World*. Harmondsworth: Penguin, 1955

James, William. *Principles of Psychology*. New York: Holt, 1890

Jane, Cecil (ed.). *The Voyages of Christopher Columbus*. London: The Argonaut Press, 1930

Jaspers, Karl. *Anselm and Nicholas of Cusa*. Edited by Hannah Arendt, translated by Ralph Manheim. New York: Harcourt, Brace and World, 1966

Jules-Rosette, Bennetta. 'The veil of objectivity', *American Anthropologist*, vol. 80, 1978, pp. 549–70

Kaplan, Abraham. *The Conduct of Inquiry*. San Francisco: Chandler, 1964

Kaufman, Felix. *The Methodology of the Social Sciences*. New York: Oxford University Press, 1944

Klotz, Irving M. 'The N-ray affair', *Scientific American*, vol. 242, no. 5, 1980, pp. 168–75

Knorr, Karen D. 'Producing and reproducing knowledge: descriptive or reconstructive', *Social Science Information*, vol. 16, no. 6, 1977, pp. 669–96

Koestler, Arthur. *The Sleepwalkers*. London: Hutchinson, 1959
  The Act of Creation. London: Hutchinson, 1964
  Ghost in the Machine. London: Hutchinson, 1967
  The Case of the Midwife Toad. London: Hutchinson, 1971

Koffka, W. 'Gestalt', pp. 642–5 in *The Encyclopedia of the Social Sciences*, vol. 3, edited by Edwin R. A. Seligman. New York: Macmillan, 1931

Köhler, W. *The Mentality of Apes*. London: Pelican Books, 1957

Kordig, Carl R. *The Justification of Scientific Change*. Dordrecht: Reidel, 1971

Kottler, Malcolm J. 'Hugo de Vries and the rediscovery of Mendel's laws', *Annals of Science*, vol. 36, 1979, pp. 517–38

Kroeber, A. L. 'The superorganic', *American Anthropologist*, vol. 19, no. 2, 1917, pp. 163–213

Kuhn, Thomas S. 'Historical structures of scientific discoveries', *Science*, vol. 136, June 1, 1962, pp. 760–4
  'The function of dogma in scientific research', pp. 59–89 in *Scientific Change*, edited by A. C. Crombie. London: Heinemann, 1963
  'The essential tension: tradition and innovation in scientific research', pp. 341–54 in *Scientific Creativity, Its Recognition and Development*, edited by Calvin W. Taylor and Frank Barron. New York: Kreiger, 1963
  The Structure of Scientific Revolutions. Second edition. Chicago: University of Chicago Press, 1970
  'Reflections on my critics', pp. 231–78 in *Criticism and the Growth of Knowledge*, edited by Imre Lakatos and Alan Musgrave. London: Cambridge University Press, 1970
  'Second thoughts on paradigms', pp. 459–82 in *The Structure of Scientific Theories*, edited by Frederick Suppe. Urbana: University of Urbana Press, 1974
  The Essential Tension: Selected Studies in Scientific Tradition and Change. Chicago: University of Chicago Press, 1977

Lakatos, Imre and Alan Musgrave (eds.). *Criticism and the Growth of Knowledge*. London: Cambridge University Press, 1970

Lanczos, Cornelius. *The Einstein Decade*. London: Elek Science, 1974

Latour, Bruno and Steve Woolgar. *Laboratory Life: The Social Construction of Scientific Facts.* Beverly Hills: Sage, 1979

Laudan, Larry. *Progress and its Problems.* Los Angeles: University of California Press, 1977

Law, John and David French. 'Normative and interpretive sociologies of science', *Sociological Review,* vol. 22, 1974, pp. 581–95

Lecourt, D. *Proletarian Science.* Translated by Ben Brewster. London: New Left Books, 1977

Lee, K. K. 'Popper's falsifiability and Darwin's natural selection', *Philosophy,* vol. 44, no. 170, 1969, pp. 291–302

Leiss, William. *The Domination of Nature.* New York: George Braziller Publishing, 1972

Lowes, John L. *The Road to Xanadu.* New York: Vintage Books, 1959

Mackay, Robert W. 'The ethnography of the classroom', unpublished Ph.D. dissertation, Department of Sociology, University of California at Santa Barbara, 1974

 'Conceptions of children and models of socialization', pp. 180–93 in *Ethnomethodology,* edited by Roy Turner. Harmondsworth: Penguin, 1974

 'Standardized tests: objective and objectified measures of competence', pp. 218–47 in *Language Use and School Performance,* edited by A. V. Cicourel *et al.* New York: Academic Press, 1974

Madariaga, Salvador de. 'Christopher Columbus', pp. 937–42 in *Encyclopaedia Britannica,* vol. 4, fifteenth edition. Chicago: H. H. Benton, 1974

Mannheim, Karl. *Ideology and Utopia.* Translated by Louis Wirth and Edward A. Shils. New York: Harcourt, Brace and World, 1936

McKinney, John C. and Edward Tiryakian. *Theoretical Sociology.* New York: Appleton, Century, Crofts, 1970

Medvedev, Zhores. *The Rise and Fall of T. D. Lysenko.* Garden City: Doubleday, 1971

Mendel, Gregor. 'Experiments on plant hybrids', pp. 1–48 in *The Origin of Genetics: A Mendel Sourcebook,* edited by Curt Stern and Eva R. Sherwood. San Francisco: Freeman, 1966

Mermin, N. David. *Space and Time in Special Relativity.* New York: McGraw Hill, 1968

Merton, Robert K. *The Sociology of Science, Theoretical and Empirical Investigations.* Edited and with introductory remarks by Norman W. Storer. Chicago: University of Chicago Press, 1973

Millar, Ronald. *The Piltdown Men.* Letchworth: Victor Gollancz, 1972

Millhauser, Milton. *Just Before Darwin.* Middletown: Wesleyan University Press, 1959

Morison, Samuel Eliot. *The European Discovery of America.* New York: Oxford University Press, 1971

Mulkay, Michael J. 'Some aspects of cultural growth in the natural sciences', *Social Research,* vol. 36, 1969, pp. 22–52

 The Social Process of Innovation. London: Macmillan, 1972

 'Methodology in the sociology of science', *Social Science Information,* vol. 13, 1974, pp. 107–19

'Three models of scientific development', *Sociological Review,* vol. 23, 1975, pp. 509–25

'Norms and ideology in science', *Social Science Information,* vol. 15, 1976, pp. 637–56

'The mediating role of the scientific elite', *Social Studies of Science,* vol. 6, 1976, pp. 445–70

'Sociology of the scientific research community', pp. 93–148 in *Science, Technology and Society,* edited by I. Spiegel-Rosing and D. Price. Beverly Hills: Sage, 1977

'Consensus in science', *Social Science Information,* vol. 17, 1978, pp. 107–22

Science and the Sociology of Knowledge. London: Allen and Unwin, 1979

Mullins, Nicholas. 'The distribution of social and cultural properties in informal community networks among biological scientists', *American Sociological Review,* vol. 33, 1968, pp. 786–97

Ogburn, William F. *Social Change.* New York: Heubsch, 1922

and Dorothy Thomas. 'Are inventions inevitable? A note on social evolution', *Political Science Quarterly,* vol. 37, no. 1, 1922, pp. 83–98

O'Gorman, Edmundo. *The Invention of America.* Bloomington: Indiana University Press, 1961

Olby, Robert. *The Origins of Mendelism.* London: Constance, 1966

'Mendel no Mendelian?', *History of Science,* vol. 17, 1979, pp. 53–72

Partington, J. R. 'The discovery of oxygen', *Journal of Chemical Education,* vol. 39, no. 3, 1962, pp. 123–5

Pitkin, Hanna. *Wittgenstein and Justice.* Los Angeles: University of California Press, 1972

Plato. *The Collected Dialogues of Plato.* Edited by E. Hamilton and H. Cairns. Princeton: Princeton University Press, 1961

Poincaré, Henri. *Science and Method.* Translated by Francis Maitland. New York: Scribner, 1914

Polanyi, Michael. *Personal Knowledge.* Chicago: University of Chicago Press, 1958

The Tacit Dimension. Garden City: Doubleday, 1966

Pollner, Melvin. 'Mundane reasoning', *Philosophy of the Social Sciences,* vol. 4, 1974, pp. 35–54

'Sociological and common-sense models of the labelling process', pp. 27–41 in *Ethnomethodology,* edited by Roy Turner. Harmondsworth: Penguin, 1974

'The very coinage of your brain', *Philosophy of the Social Sciences,* vol. 5, 1975, pp. 411–30

Popper, Karl. *The Logic of Scientific Discovery.* New York: Basic Books, 1959

The Poverty of Historicism, second edition. London: Routledge and Kegan Paul, 1960

Conjectures and Refutations. New York: Basic Books, 1963

'Normal science and its dangers', pp. 51–8 in *Criticism and the Growth of Knowledge,* edited by Imre Lakatos and Alan Musgrave. London: Cambridge University Press, 1970

Price, Derek J. de Solla. *Little Science, Big Science*. New York: Columbia University Press, 1963
  'The peculiarity of a scientific civilization', pp. 1–24 in *Science Since Babylon*, enlarged edition, New Haven: Yale University Press, 1975
Reichenbach, Hans. *Experience and Prediction*. Chicago: University of Chicago Press, 1938
Roberts, H. F. *Plant Hybridization Before Mendel*. Princeton, N.J.: Princeton University Press, 1929
Rosner, Stanley and Lawrence Abt (eds). *The Creative Experience*. New York: Grossman, 1970
Rostand, Jean. *Error and Deception in Science*, translated by A. R. Pomerans. London: Hutchinson, 1960
Rupke, Nicholaas A. '*Bathybius haeckelii* and the psychology of scientific discovery', *Studies in the History and Philosophy of Science*, vol. 7, no. 1, 1976, pp. 53–62
Russett, Cynthia Eagle. *Darwin in America*. San Francisco: Freeman, 1976
Salmon, Wesley C. *Logic*. Englewood Cliffs: Prentice-Hall, 1963
Sapir, Edward. 'Do we need a superorganic?', *American Anthropologist*, vol. 19, 1917, pp. 442–3
Schutz, Alfred. *Phenomenology of the Social World* (1932). Translated by George Walsh and Frederick Lehnert. Evanston: Northwestern University Press, 1967
  The Collected Papers, vol. 1: *The Problem of Social Reality*. Edited by Maurice Natanson. The Hague: Martinus Nijhoff, 1962
Selye, Hans. *From Dream to Discovery*. New York: McGraw Hill, 1964
Shapere, Dudley. 'The structure of scientific revolutions', *Philosophical Review*, vol. 73, 1964, pp. 383–94
  'The paradigm concept', *Science*, vol. 172, 1971, pp. 706–9
Shor, Elizabeth N. *The Fossil Feud Between E. D. Cope and O. C. Marsh*. New York: Exposition Press, 1974
Simon, Herbert A. *Models of Discovery and Other Topics in the Methods of Science*. Dordrecht: Reidel, 1977
Sinclair, W. J. *Semmelweis: His Life and His Doctrine*. Manchester: Manchester University Press, 1909
Stent, Gunther S. *Paradoxes of Progress*. San Francisco: Freeman, 1978
Stephens, James. *Francis Bacon and the Style of Science*. Chicago: University of Chicago Press, 1975
Stern, Curt and Eva R. Sherwood (eds.). *The Origin of Genetics: A Mendel Sourcebook*. San Francisco: Freeman, 1966
Stuckey, William K. 'The prize', *The Saturday Review*, vol. 55, no. 36, 1972, pp. 33–9
Suppe, Frederick (ed.). *The Structure of Scientific Theories*. Urbana: University of Urbana Press, 1974
Taton, René. *Reason and Chance in Scientific Discovery*. London: Hutchinson, 1957
Taylor, Calvin W. and Frank Barron (eds.). *Scientific Creativity, Its Recognition and Development*. New York: Kreiger, 1963

Taylor, L. W. *Physics, the Pioneer Science*. Boston: Houghton Mifflin, 1941

Thacker, John Boyd. *Christopher Columbus*, vol. 2 (1902). New York: AMS Press, 1967

Turner, Roy. 'The ethnography of experiment', *American Behavioral Scientist*, vol. 10, no. 8, 1967, pp. 26–9

    'Words, utterances and activities', pp. 169–87 in *Understanding Everyday Life*, edited by Jack D. Douglas. Chicago: Aldine, 1970

    (ed.). *Ethnomethodology*. Harmondsworth: Penguin, 1974

Wade, Nicholas. 'I.Q. and heredity: suspicion of fraud beclouds classic experiment', *Science*, vol. 194, November 20, 1976, pp. 916–19

    'Voice from the dead names new suspect for Piltdown hoax', *Science*, vol. 202, December 8, 1978, p. 1062

Wertheimer, Max. 'On truth', *Social Research*, vol. 1, 1934, pp. 135–46

    'Gestalt theory', *Social Research*, vol. 11, 1944, pp. 78–99

    Productive Thinking. New York: Harper, 1945

White, Leslie. *The Science of Culture*, second edition. New York: Farrar, Straus and Giroux, 1969

Winch, Peter. *The Idea of a Social Science and its Relation to Philosophy*. London: Routledge and Kegan Paul, 1958

Wittgenstein, L. *The Blue and Brown Books*. Oxford: Basil Blackwell, 1958

Woolgar, Steve. 'Writing an intellectual history of scientific development: the use of discovery accounts', *Social Studies of Science*, vol. 6, 1976, pp. 395–422

    'Discovery: logic and sequence in a scientific text', in *Sociology of the Sciences Yearbook*, vol. 5: *The Social Process of Scientific Investigation*. Dordrecht: Reidel, 1981 forthcoming

Zamiatin, E. I. *We* (1924). Translated by B. G. Iverney. London: Jonathan Cape, 1970

Ziman, John. *Reliable Knowledge*. Cambridge: Cambridge University Press, 1978

Zirkle, Conway. 'Gregor Mendel and his precursors', *Isis*, vol. 42, 1951, pp. 97–104

Zuckerman, Harriet. 'Nobel laureates in science', *American Sociological Review*, vol. 32, 1967, pp. 391–403

    'The Sociology of the Nobel prize', *Scientific American*, vol. 217, no. 5, 1967, pp. 23–33

    'Patterns of name-ordering among authors of scientific papers', *American Journal of Sociology*, vol. 74, no. 4, 1968, pp. 276–91

    The Scientific Elite. New York: Free Press, 1977

Index